Photovoltaics in Cold Climates

Photovoltaics in Cold Climates

Technical editors:
MICHAEL ROSS
CANMET Energy Diversification Research Laboratory
and
JIMMY ROYER
Solener Inc.

First published 1999 by James & James (Science Publishers) Ltd

This edition published 2013 by Earthscan

For a full list of publications please contact:
Earthscan
2 Park Square, Milton Park, Abingdon, Oxon OX14 4RN
Simultaneously published in the USA and Canada by Earthscan
711 Third Avenue, New York, NY 10017

Earthscan is an imprint of the Taylor & Francis Group, an informa business

A catalogue record for this book is available from the British Library

ISBN 978-1-873936-89-4 (pbk)

Illustrations
Plam de Quebec

Cover photographs
Mountaintop telecommunication facility: Jim Allen, NorthwesTel Inc.
Parks Canada facility: Michael Ross, CANMET EDRL

Disclaimer
This book was written for informational purposes and does not necessarily reflect the views of the Government of Canada, the International Energy Agency, or the employers of any authors. Neither Canada nor the International Energy Agency nor any of their ministers, officers, employees or agents makes any warranty in respect to this book or assumes any liability arising out of this book

Contents

Preface/ Préface

There are many excellent books which describe the design and operation of photovoltaic (PV) systems. The majority, however, were written for warm climates. There are important differences between systems in these climates and those found in cold regions, and this book is dedicated to those differences; it aims to complement, rather than replace, existing books.

This book is intended to satisfy the needs of a varied audience, but is primarily written for decision-makers – the people who will decide, for a given application, whether PVs will be considered as an alternative to traditional power sources. For the decision-makers, we have tried to demonstrate that PV is a viable option in cold climates, while drawing attention to some of the issues that arise because of climate. In the first section of the book, technical detail has been avoided; such detail can be found in the works referenced at the end of each chapter.

While the decision-maker's needs have been given priority, the engineer and user have not been ignored. Practical considerations dominate the second, third and fourth sections of the book. These sections have been written assuming that the reader has some knowledge about photovoltaics.

Lastly, the book shows different examples of successful PV applications in cold climates around the world. In most of these examples, photovoltaics were chosen because they were considered to be the power source which was the most economical and practical to implement.

This book, being a collaboration, owes its existence to many people and organizations. Two programmes were instrumental to its realisation: the Canadian *Photovoltaics for the North* programme and the International Energy Agency's *Photovoltaic Power Systems* (PVPS) Programme.

Photovoltaics for the North is a joint undertaking of the CANMET Energy Diversification Research Laboratory of Natural Resources Canada, the Aurora Research Institute (Inuvik, Northwest Territories) and the Nunavut Research Institute (Iqaluit, Northwest Territories). It is aimed at increasing the penetration of photovoltaic technologies in Canada's North through technology development and transfer activities. It is anticipated that this book will serve the needs of decision-makers, engineers and PV system users in northern Canada long after the completion of the *PV for the North* programme. *PV for the North* provided the impetus and funding for this book.

Plusieurs excellents livres décrivent la conception et le fonctionnement des systèmes photovoltaïques (PV). Toutefois, la majorité d'entre eux porte sur le fonctionnement de ces systèmes sous un climat chaud. Il existe d'importantes différences entre l'utilisation d'un système PV sous ce climat et l'utilisation du même système sous les climats froids. Le présent ouvrage étudie ces différences. Il a pour but de compléter plutôt que de remplacer les ouvrages existants.

Le présent ouvrage s'adresse à un public diversifié, mais plus spécialement aux décideurs – ceux qui vont décider si on peut envisager la possibilité d'avoir recours aux systèmes PV, pour une application donnée, comme solution de rechange aux systèmes énergétiques traditionnels. Nous nous sommes efforcés de démontrer aux décideurs que le système PV est une option viable sous les climat froids tout en soulignant certains des problèmes auxquels il faut faire face en raison du climat. Dans la première partie de l'ouvrage, nous avons évité de donner des détails trop techniques. Pour les obtenir, le lecteur consultera les ouvrages de référence cités à la fin de chaque chapitre.

Nous avons donné la priorité au décideur, sans toute fois négliger les aspects techniques et pratiques. La deuxième, la troisième et la quatrième parties de l'ouvrage sont davantage axées sur des considérations pratiques. Nous avons présumé que le lecteur a une certaine connaissance des systèmes PV.

Finalement, l'ouvrage donne différents exemples d'applications réussies de systèmes PV sous les climats froids partout dans le monde. Dans la plupart de ces exemples, les systèmes PV ont été choisis parce qu'ils s'agissait du système énergétique le plus économique et le plus facile à mettre en oeuvre.

Le présent ouvrage est le fruit de la collaboration de nombreuses personnes et de nombreux organismes. Deux programmes, en particulier, ont permis la réalisation de ce livre: le programme canadien *PV pour le Nord* et le programme *Photovoltaic Power Systems (PVPS)* de l'Agence internationale de l'énergie.

Le programme *PV pour le Nord* a été mis sur pied par le Laboratoire de recherche en diversification énergétique de CANMET de Ressources naturelles Canada, l'Aurora Research Institute (Inuvik, Territoires du Nord-Ouest) et le Nunavut Research Institute (Iqaluit, Territoires du Nord-Ouest). Le programme vise à accroître la pénétration des technologies PV sur le marché du grand Nord canadien grâce au développement et au transfert de technologies.

The International Energy Agency (IEA), founded in 1974, is an autonomous body within the framework of the Organization for Economic Cooperation and Development (OECD), which carries out a comprehensive programme of energy cooperation among its 23 member countries. The European Commission also participates in the work of the Agency.

The IEA PVPS Programme is one of the collaborative R&D agreements established within the IEA, and since 1993 its participants have been conducting a variety of joint projects in the applications of photovoltaic conversion of solar energy into electricity. The 22 members of the programme are Australia, Austria, Canada, Denmark, the European Commission, Finland, France, Germany, Israel, Italy, Japan, Korea, Mexico, The Netherlands, Norway, Portugal, Spain, Sweden, Switzerland, Turkey, the UK and the USA.

Within the IEA PVPS programme there are presently seven subgroups. The objective of the first subgroup, known as 'Task 1', is to promote and facilitate the exchange and dissemination of information on the technical, economic and environmental aspects of photovoltaics.

We would like to acknowledge the important contributions of all the authors. While some of these experts represented their countries in the IEA PVPS Task 1, most of them contributed their time and knowledge because they believed that this book was needed to expand understanding and use of photovoltaics in cold climates.

A number of people have worked behind the scenes in the preparation of the book. We would like to thank Harry Barnes for his commitment and tireless support as former Operating Agent of Task 1, Sylvain Martel for project management and many helpful suggestions, Lisa Dignard-Bailey and Greg Leng for suggesting further improvements, Dave Turcotte and Govindasamy Tamizhmani for their technical expertise, André Filion for his continued support and Edward Milford for his guidance and patience. Finally, we would like to thank Raye Thomas and Eric Usher for initiating this project.

MICHAEL M. D. ROSS
JIMMY ROYER

Nous pensons que le présent ouvrage continuera à répondre aux besoins des décideurs, des techniciens et des utilisateurs de systèmes PV dans le nord du Canada longtemps après que le *Programme PV pour le Nord* aura pris fin. Ce programme a assuré le financement de l'ouvrage et a donné l'impulsion nécessaire à sa réalisation.

L'Agence internationale de l'énergie (AIE), fondée en 1974, est un organisme autonome de l'Organisation de coopération et de développement économiques (OCDE). L'Agence a un programme complet de coopération en matière d'énergie entre ses 23 pays membres. La Commission européenne collabore également à l'Agence.

Le programme PVPS de l'AIE a été mis sur pied à la suite de l'une des ententes de collaboration en matière de recherche et développement conclues au sein de l'AIE. Depuis 1993, ses participants mènent à bien divers projets communs en ce qui concerne les applications de la conversion photovoltaïque de l'énergie solaire en électricité. Les 22 pays qui participent au programme sont l'Allemagne, l'Australie, l'Autriche, le Canada, la Commission européenne, la Corée, le Danemark, l'Espagne, les États-Unis d'Amérique, la Finlande, la France, Israël, l'Italie, le Japon, le Mexique, la Norvège, les Pays-Bas, le Portugal, le Royaume-Uni, la Suède, la Suisse, et la Turquie.

Le programme PVPS compte actuellement sept groupes. Le premier sous-groupe que l'on appelle 'Tâche 1' a pour objectif de promouvoir et de faciliter l'échange et la diffusion de l'information sur les aspects techniques, économiques et environnementaux de la technologie photovoltaïque. La majeure partie des recherches faites pour le présent ouvrage ont été effectuées dans le groupe de la Tâche 1.

Nous aimerions remercier tous les auteurs pour leur importante contribution. Certains des experts représentaient leurs pays au groupe de la Tâche 1; la plupart d'entre eux ont communiqué leurs connaissances et donné de leur temps parce qu'ils jugeaient nécessaire la publication du présent ouvrage pour mieux faire comprendre et promouvoir l'utilisation des technologies PV sous les climats froids.

Un certain nombre de personnes ont collaboré au présent ouvrage. Nous tenons à remercier Harry Barnes pour son dévouement et son soutien indéfectible à titre de responsable du groupe de la Tâche 1, Sylvain Martel pour la gestion du projet et ses nombreuses et utiles suggestions, Lisa Dignard-Bailey et Greg Leng pour avoir proposé plusieurs améliorations, Dave Turcotte et Govindasamy Tamizhmani pour leur expertise technique, André Filion pour son soutien inconditionnel et Edward Milford pour sa patience et ses conseils. Finalement, nous remercions Raye Thomas et Eric Usher pour avoir initié le projet.

MICHAEL M. D. ROSS
JIMMY ROYER

I General

1 Introduction

Michael M.D. Ross
(CANMET Energy Diversification Research Laboratory, Natural Resources Canada)

For many people, solar energy is associated exclusively with warm, sunny climates; many more see it as a power source for the future. Few people are aware that thousands of solar systems have already been installed in those parts of the world that experience snow, ice and bone-chilling temperatures.

At first glance, cold climates and solar power may seem incompatible. This is a misconception: not only do solar systems work at cold temperatures, but they are also the least expensive source of electricity for many different applications.

This book focuses on photovoltaics (PVs), the solar technology that turns light directly into electricity. Photovoltaics need sunlight, not heat, to generate power. Cold-climate regions are by definition cold, but they are not necessarily dark. During winter, cold regions do receive less energy from the sun than more equatorial locales. Fortunately, the converse is also true: during summer, PV systems in places like Alaska and Lapland can generate more electrical energy than similar systems at the equator.

A large market for photovoltaics already exists and in the future this market will expand as environmental concerns lead to increased demand for non-polluting power sources like photovoltaics. In the meantime, PVs are used at many sites not reached by the electric 'grid', that vast web of power lines and central power plants that supplies electricity to most people in the industrialized world. Photovoltaics are simpler, more reliable and cheaper than other power sources and, although they are expensive to purchase, they cost next to nothing to operate – sunshine is free and maintenance requirements are minimal. For modest power requirements, PVs are generally the most cost-effective option.

WHAT IS PHOTOVOLTAIC ENERGY?

As the name would suggest, solar power systems convert sunshine into useful power. One solar power technology is the *photovoltaic cell*, a specially prepared semiconductor that is sensitive to sunlight. Expose a cell to light and it generates electricity. A single photovoltaic cell generates very little power, but a number of cells can be assembled into a module, in which cells are sandwiched between a flat sheet of glass and a protective plastic backing. The module is a convenient package capable of providing a reasonable amount of power; for higher power, modules are wired together to form an array.

If the photovoltaic array is wired directly to an electric load, the load can operate only when there is light. The addition of a storage battery permits the system to operate at any time. The array charges the battery on sunny days and the battery powers the load at night and on cloudy days.

Examination of the cell on a solar-powered calculator reveals the advantages of photovoltaic technology. First, no moving parts, which have a tendency to break, are involved. As a result, PV cells can operate without maintenance for 20 years or more. Second, light is the 'fuel'. A reliable supply of sunshine has been freely available for the last five billion years and, regardless of inflation, this is not expected to change any time soon. Third, it is a very simple technology. Shine light on the cell and the calculator turns on. There are no complicated operating procedures and very little can go wrong. And fourth, though it is not evident from the calculator, photovoltaics are modular. Adding and removing modules from an array is straightforward, so it can expand or shrink in response to changing requirements.

Of course, photovoltaic technology has weaknesses, too. It is expensive; when electricity is available from the grid or much electricity is needed, PVs are rarely cost-effective. On the other hand, the price of PVs per unit of power-generating capacity has plummeted over the past 20 years and will continue to fall as the technology advances and manufacturing processes improve (Figure 1.1). Another problem is PV technology's reliance on an intermittent resource: modules only generate electricity when the sun shines, so batteries are usually necessary if PVs are the only source of power. In addition, the 'power density' of sunlight is relatively low: a PV array must be large to generate much power. Finally, economies of scale, though significant in the manufacture of PV modules, do not greatly reduce installation costs.

WHAT IS A COLD CLIMATE?

'Cold' is a relative term and 'cold climate' has different connotations for different people. For the purposes of this

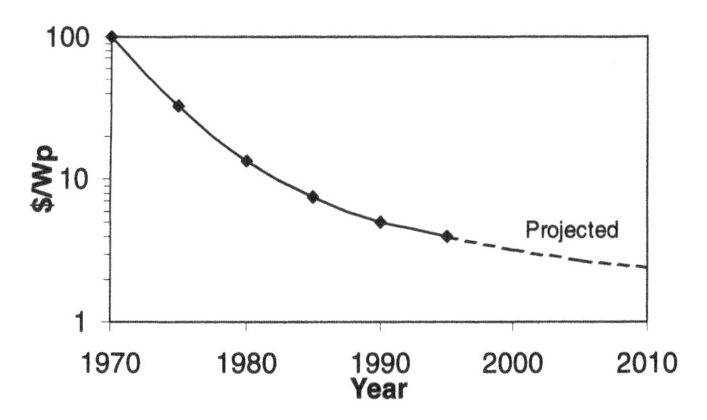

Figure 1.1. The price of photovoltaics, past and projected[1]

book, a cold climate is defined as one in which snow, ice, freezing temperatures, dark winters and long summer days are significant considerations in the design and operation of a PV system.

The climates that fit this description tend to occur at middle and high latitudes and at high altitudes. High-latitude cold-climate areas include the Antarctic, the Arctic, Scandinavia, northern Russia, Alaska, much of Canada and parts of the northern USA, Central Asia and Europe. Mountainous areas are notoriously cold and PV systems located in, for example, the Andes or the European Alps face many of the climatic conditions associated with high-latitude regions.

COLD-CLIMATE APPLICATIONS OF PV

The types of applications to which photovoltaic technology is well suited arise frequently in cold climates. Cold regions are harsh places that tend to be sparsely populated and somewhat inaccessible. As a result, they typically have a low concentration of infrastructure. It is not difficult to find a spot where the nearest power line is kilometres, or even hundreds of kilometres, away. Extending the power line to such remote sites is often prohibitively expensive; likewise, the cost of servicing these sites is high, because the necessary expertise and supplies must be transported at great cost from distant population centres. While it may seem strange that a device requiring electricity would be so remotely located, such applications are surprisingly common: they include telecommunication repeater stations, monitoring systems, homes and cottages, oil and natural gas wells, mountain lodges, research stations, hunting and fishing camps, and even remote communities (see International Energy Agency[2] for more examples).

When reliability is paramount: telecommunications and monitoring

Photovoltaic power is used for numerous cold-climate telecommunication and monitoring systems. For these systems, reliability and low maintenance requirements are paramount: a power outage can mean the loss of valuable data or

Figure 1.2. A telecommunications repeater station powered by a PV–hybrid system; the site is in Nahanni National Park, Northwest Territories, Canada (62°N) (*photo:* NorthwesTel Inc.)

the inability to communicate. Often they must operate unattended and maintenance visits can be expensive. These requirements make PV an obvious choice.

Telecommunications repeater stations constitute a major cold-region market for PV technology. In many high-latitude areas, radio-frequency and microwave repeater stations form the backbone of the telecommunications system. Signals, such as telephone calls, are relayed from repeater station to repeater station. To maximize the range of the signal and the area covered, repeater stations are often situated along ridges and on mountains, where the climate can be extreme.

Each repeater station requires a reliable source of electricity. Being located on mountain tops and in areas of low population density, the stations are often difficult to access and occasionally can be reached only by helicopter. Assuming that the grid does not extend to the site, the options for powering the repeater station are limited. In the past, common solutions were fossil fuel-powered generators or non-rechargeable batteries. Both required frequent visits for servicing and/or maintenance; this raised operating costs. PV systems, on the other hand, offer high reliability and very low operating costs and are therefore widely used for repeater stations (Figure 1.2).

The amount of power required by the telecommunications equipment varies from one repeater station to the next. For low-power systems, such as VHF radio towers, PV systems can have payback periods as short as one year when compared with primary batteries. In a study of VHF repeaters in Canada's Northwest Territories, the value of savings over the life of the repeater station was five times larger than the initial cost of the PV installation.[3]

At repeater stations requiring more power, it may not make sense to use PVs alone. Rather, it may be more cost-effective to combine them with a fossil fuel-powered generator in a 'hybrid' power system. The PV array powers the load and charges a battery when sunshine is abundant; the fossil fuel-powered generator runs when neither the array nor the battery can power the load. Reduced fuel consumption can justify the additional expense of the PV system, especially at those installations at which fuel delivery costs are high. Since the generator can be used when loads are large or little solar energy is available, the PV system does not have to be oversized in anticipation of these relatively rare occurrences; this lowers the capital cost of the PV system. Additional benefits of PV–hybrid systems are lower maintenance costs due to the reduced load on the generator and increased reliability stemming from redundancy of power sources.

A PV system can be considered expensive only if there is a cheaper alternative. For example, a string of repeater stations relay communications between Alert (82°30′N) and Eureka (80°00′N), on Ellesmere Island, the northernmost island of significance in the Canadian polar archipelago. Polar night descends on these telecommunications systems between November and February, yet they are powered by PV. Presently, PV supplies power from March through October and a primary battery is used for the remainder of the year. Replacing these batteries – which involves flying by helicopter to each site every second year – is extremely expensive, however, and a planned expansion of the PV system will render them unnecessary. In this new system, a very large rechargeable battery bank will power the station during the winter and be recharged by the PV system during the summer. Average temperatures during the winter approach –40°C and a thermal management system will be required. The system will be large and expensive, but will be the most cost-effective solution capable of satisfying the constraints on reliability.

Many natural phenomena and industrial processes are monitored and controlled: meteorological stations track weather conditions, hydrological instruments gather information on water levels in reservoirs and flow meters measure and control the production of natural gas at well-heads. Many such monitoring systems are situated in cold-climate regions, often far from convenient power sources. While the equipment does not require much electricity, the supply must be reliable, especially when the system is being relied on to control a crucial process.

The rolling prairie around Calgary, Alberta, Canada is dotted with oil and gas well-heads. Each well-head is monitored and some are remotely controlled; the power supply must be highly reliable, since important data is lost if the monitoring system shuts down. Because there are thousands of wells, maintenance requirements for the power system must be minimal, or the cost of servicing the installations would be prohibitively high. Winter temperatures drop below –30°C, but it is also relatively bright, with clear skies being the norm, and photovoltaics are often the least-cost solution (Figure 1.3).

Although many of the well-heads are located very close to power lines, photovoltaics are still used. The cost of the connection to the electric grid, the hassle of occasional grid failures and the advantages of having a standard system all work in photovoltaics' favour.

Transportability: navigational buoys and laptop computers

Certain devices must be transportable; wherever they are situated, they must function, without external power or intervention. PVs are ideally suited to these devices, owing to their reliability, simplicity and the near universal availability of natural light. As demonstrated by the proliferation of

Figure 1.3. A PV power system for monitoring equipment at a gas well near Calgary, Alberta, Canada. Located at 51°N, winter temperatures reach –40°C. (*photo:* Soltek Solar Energy Ltd.)

Figure 1.4. Navigational buoy in the Atlantic Ocean near the coast of The Netherlands (*photo:* Shell Solar Energy BV)

solar-powered calculators and watches, this is especially true when the power requirement is relatively small.

The Canadian Coastguard deploys approximately 6000 lighted navigational buoys powered by PVs. Such buoys operate year-round on both the Atlantic and Pacific coasts. Obviously, connecting to the electrical grid is not feasible and replacing multitudes of non-rechargeable batteries, spread along thousands of kilometres of coastline, would be extremely expensive. A PV module and a rechargeable battery have become the standard solution in Canada and many other coastal countries of the world, despite the harsh, saline environment (Figure 1.4).

Less crucial applications can also benefit from portable PV power. Geologists, biologists and other researchers who set up field camps usually need a small source of power to operate laptop computers and other electronic devices; in these camps, the weight of equipment can be a major concern. For small power supplies, PVs perform well and have become quite popular among the scientists who venture into the Canadian Arctic each summer (Figure 1.5).

Simplicity: remote homes and schools

The simplicity of photovoltaic systems permits them to be used in those remote locations where complex generators would break and no one would be able to repair them. Little expertise is required to maintain a photovoltaic system and little maintenance is required.

Figure 1.5 Flexible solar module powering a laptop computer near Yellowknife, Northwest Territories, Canada (63°N) (*photo:* Midnight Sun Energy)

It is estimated that two billion people, the majority in developing countries, have no access to electricity. Many of these people would appreciate the convenience of electricity. Photovoltaics are a viable solution, a fact underlined by the millions of dollars that the World Bank and other organizations have committed to PV rural electrification. While most of these funds target tropical regions, they need not exclude cold-climate areas, including Western and Northern China, Nepal, Afghanistan, Mongolia and the South American Andes. The electricity provided by a PV module can significantly affect rural populations in developing countries: it permits work and study at night; enables operation of equipment necessary for small enterprises; facilitates communication with the rest of the world by telephone, radio and television; and fosters a sense of modernity to counteract the lure of overcrowded cities.

PVs have been successfully applied in rural electrification projects for cold-climate developing countries. On the steppes of Mongolia, where average January temperatures are –25°C, PVs have provided power for the homes of nomadic herdspeople (Figure 1.6). Living in portable tents, the herdspeople frequently move themselves and their belongings; the PV systems can be dismantled and reassembled to accommodate these moves. In the Argentinean Andes, remote school-houses have been equipped with PV systems that power lights and small electric appliances. The teachers and caretakers have been given basic training to help them understand how the PV system works and perform simple maintenance tasks. During winter, the school-houses are literally buried in snow and classes do not run, but the systems have operated well nevertheless: snow tends to melt or slide off PV modules quite reliably (see Case Studies).

It is not just people in developing countries who live 'off-grid'. In North America, a significant number of people choose to live without the grid, for economic or personal reasons. For example, in Yellowknife, the capital of the Canadian Northwest Territories, restrictions on construction outside city limits inflate land costs and consequently houseboats have become popular on adjacent Great Slave Lake. PVs provide electricity to about 20% of these houseboats, often as part of a hybrid system.

Summer-only applications: cottages and lodges

Cold temperatures and snow affect humans as well as power systems. For this reason, certain activities are associated nearly exclusively with the warmer months of the year. Since these activities coincide with a period of abundant sunshine, PVs are well suited to any power requirements they may have.

In addition to remote residences that are occupied throughout the year, there are a vast number of summer cottages that rely on PV for electricity (Figure 1.7). These systems are common in Canada and Scandinavia; Norway alone has

Figure 1.6. Portable system for Mongolian herdspeople; January temperatures average below –25°C (*photo:* IT Power)

Figure 1.7. A summer cottage in Finland: note the two PV modules under the upper balcony (*photo:* Neste Advanced Power Systems)

60,000 PV-powered cottages. Typically, these PV systems are very small – a few hundred watts at the most – but coupled with a battery and efficient energy use, they provide enough electricity to satisfy the cottage-dweller's needs. This is due to the good match between solar availability and cottage use: both are at their maximum during summer. In addition, long summer days minimize heating and lighting requirements.

Tourism usually peaks during the summer, especially in cold-climate regions. Lodges, campgrounds, alpine huts and other off-grid tourist destinations can use PVs. Many remotely situated hunting and fishing lodges are powered principally by PVs. For extended cloudy periods, times of peak loads and late autumn use, most of these lodges also have a gasoline, diesel or propane generator.

When intermittent power is acceptable: fans and pumps

Some devices do not need power at all times: intermittent power is acceptable and the battery can be eliminated from a PV system, making it cheaper and more reliable. Fans and pumps are such devices.

In many Arctic communities, there is too much sun during summer and keeping a house cool is problematic. Fans are often placed in the attics of existing buildings. Installing a standard fan involves extending the house wiring and adding a thermostat and controller. Many people have found it simpler and cheaper to wire a fan directly to a PV module. When the sun shines, the attic warms up and the fan turns.

PV-powered water pumps have been used in tropical and desert countries for many years. More recently, they have been introduced to livestock watering stations in colder regions, such as the Canadian prairies. When water is pumped from a watering hole to a trough, livestock do not have to enter the slough in order to drink, the slough does not become polluted and livestock stay healthy. Most watering holes are found at a distance from the electric grid, so PVs and wind power are the least-cost power sources. In such systems, the water trough acts as its own storage and no battery is necessary.

Noise- and pollution-sensitive environments: remote parks

In some environments, noise and pollution are especially undesirable. Visitors to a wilderness park, for example, generally do not appreciate the drone of a generator, the sight of rusting fuel barrels or a vista bisected by a power line. The plants and animals that reside in wilderness parks welcome pollution even less; in fragile Arctic environments, for example, even a minor fuel spill can be extremely damaging. In such environments, silent and pollution-free PV power is an excellent alternative to grid extension, generators and non-rechargeable batteries.

Figure 1.8. The PV array for a hybrid system at the Warden Station in Ellesmere Island National Park, Canadian Arctic

As an example of such an environment, consider Ellesmere Island National Park, high in the Canadian Arctic. The park wardens stationed at Tanquary Fiord (81°N) rely on PVs for about half their electricity, with the remainder coming from a wind turbine and a diesel generator. The availability of sunlight matches the electrical requirement very well: the park is 'open' during the summer only – polar night creeps over the park in autumn. All fuel for the generator must be flown to Tanquary Fiord from Resolute, itself a remote community, 800 km away; fuel is, therefore, very expensive. Yet the park wardens appreciate photovoltaics not for their considerable cost savings, but rather because they are quiet and do not harm the environment the wardens are trying to protect (Figure 1.8).

Reducing generator fuel consumption: remote communities

When a lot of power is needed, PVs have difficulty competing with the cheap electricity available from the grid. Remote communities, on the other hand, are not connected to the main grid and must generate their own electricity, making PVs more attractive. For example, none of the 63 communities of the Canadian Northwest Territories is connected to the North American grid. Instead, each community or group of communities has a local grid, with electricity supplied by hydroelectric plants or, more commonly, diesel generators. Diesel fuel costs are high because of the remoteness of the communities. Consequently, electricity costs ten to twenty times more than in southern Canada. While PVs are not yet competitive in these communities, they will become so as their price drops over the coming decades (Figure 1.9).

PVs will never replace the diesel generators on these community grids; there will always be night, winter, cloudy days and other dark periods. PVs can significantly reduce the consumption of diesel fuel, however. Thus, when the sun is shining, PVs could provide a portion of the community's power, with the deficit made up by the generators. This is potentially a large market for PVs. Were they to generate just 10% of the electricity for the Northwest Territories, about 20 MW of PVs would be required – six times the total installed PV capacity in all of Canada at the end of 1997.

Clean power for everybody: grid-connected photovoltaics

Some PV systems are not connected directly to a load, but instead feed power to the electrical grid. The array behaves just like a nuclear power plant or a hydroelectric dam, except that it usually generates less power, operates only during the day and does not produce radioactive waste or cause flooding.

PVs are still too expensive to compete with traditional large-scale generation technologies on the basis of cost. As the price of PVs falls and as the environmental costs of

Figure 1.9. The PV array powering the diesel grid in Iqaluit, Northwest Territories, Canada; the array is discreetly integrated into the left-hand section of the facade

Figure 1.10. Norwegian house with a grid-tied photovoltaic array (*photo:* Norges teknisk-naturvitenskapelige unifersitet (NTNU))

existing power plants become more apparent, incorporating PVs into the grid will become increasingly attractive (Figure 1.10).

CONSIDERATIONS FOR COLD CLIMATES, CHAPTER BY CHAPTER

It has been demonstrated that PV systems work in cold climates as well as warm climates; it is important to recognize, however, that cold-climate PV systems function differently from their more tropical counterparts. Cold temperatures affect most components of a PV system. Furthermore, at high latitudes, greater variation in the amount of sunshine available during different seasons leads to systems with bigger storage systems and larger arrays. Snow and ice accumulation can also affect system design in certain installations. Following a brief introduction to PV systems (Chapter 2), equally applicable to warm and cold climates and intended for those not familiar with this technology, this book identifies and describes in detail the unique characteristics of cold-climate photovoltaics.

At locations on the equator, the sun rises at 6:00 a.m., passes more or less overhead at noon and then sets at 6:00 p.m., regardless of the day of the year. Off the equator, days are longer during summer and shorter during winter – a situation which is exacerbated at increasing distance from the equator. Thus, in high-latitude cold-climate regions, seasonal differences in the length of the solar day and the amount of solar energy available are very important, as discussed in Chapter 3.

The economic methods used to assess whether a PV system is cost-effective are generally the same in cold and warm climates. Within cold climates there are some special considerations, however; larger systems tend to be used, leading to inflated capital costs for PV systems. As pointed out in Chapter 4, this must be balanced against the fuel and maintenance costs for competing technologies, which also tend to be high.

Cold temperatures and snow and ice accumulation affect PV modules, electronic devices used for control and power conditioning and support structures (Chapter 5). Surprisingly, some of the effects are very positive: most PV modules *actually* become more efficient as their temperature drops!

The battery is the part of the PV system most severely affected by cold temperatures (Chapter 6). The electrochemical reactions at the heart of all rechargeable batteries are hampered at low temperatures, decreasing the amount of energy that can be usefully stored in the battery.

Winter's short days and cold temperatures mean that the supply of solar energy and the capacity of the battery are reduced simultaneously. This has serious implications for photovoltaic charging of batteries (Chapter 7). Bigger battery banks are necessary, so charging and discharging tend to take longer; additionally, temperature affects the ability of the battery to accept charge and control devices must account for this.

The effects of cold temperatures on PV system components affect the design of stand-alone PV systems (Chapter 8), in which PVs are the sole source of power, and PV–hybrid systems (Chapter 9), in which PVs are coupled with a generator, a wind turbine or some other source of power. In PV–hybrid systems, cold temperatures will affect not only the PV systems, but the other power sources as well.

In high-latitude PV systems, there tends to be a surplus of sunshine in the summer and a deficit during the winter. As a result, the battery must transfer energy between seasons and not just between day and night or sunny and cloudy periods (Chapter 10). In extreme cases, non-rechargeable batteries are used in conjunction with PVs; in the future, energy may be stored in the form of hydrogen, produced by a PV-powered electrolyser during the summer and feeding a fuel cell during the winter.

When determining the appropriate design for a PV system or when studying how a PV system would operate under a particular set of conditions, computerized simulation tools are often useful. The majority of these tools, however, were created with reasonably warm climates in mind and rely on assumptions and models that may not be entirely applicable in cold climates (Chapter 11). The user of such software should be aware of these shortcomings and should understand how they will affect the accuracy of the simulation.

Installing a PV system in a cold climate can be challenging. For example, during colder times of the year, using one's bare hands can be uncomfortable or even dangerous. During summer, insects such as mosquitoes and black flies can be bothersome. Practical considerations such as these are essential to a successful installation (Chapter 12).

PVs need little maintenance, but are not maintenance free, especially when batteries are incorporated into the system. Failing to maintain the system may detrimentally affect system operation and lifetime (Chapter 13).

When monitoring a PV system in a cold climate, the impact of the climate on the sensors and data acquisition system must be accounted for (Chapter 14). Snow can accumulate on pyranometers, which measure the amount of sunlight on a system, and equipment must be able to operate at cold temperatures.

The book concludes with a section of case studies illustrating a number of existing cold-climate PV systems. These cover a range of applications, demonstrating that PVs are both viable and cost-effective in cold regions.

REFERENCES

1. Maycock P (1996). *Photovoltaic Technology: Performance, Cost, and Market Forecast (1975–2010)*, fifth edition. PV Energy Systems, Inc., Catlett, VA.
2. International Energy Agency PVPS Task III (1995). *Examples of Stand-Alone Photovoltaic Systems*. London: James & James (Science Publishers) Ltd.
3. Martel S (1994). *Technico-Economic Market Study of Photovoltaic Applications in the Northwest Territories*. Varennes, Québec: Natural Resources Canada.

2 A brief introduction to photovoltaic technology

Jimmy Royer
(Solener Inc.)

Photovoltaic technology is simple, easy to understand and easy to use. For the benefit of those readers who are not familiar with photovoltaics, this chapter introduces the basic terms and concepts that are used in the rest of this book.

PHOTOVOLTAIC TECHNOLOGY

By definition, photovoltaic (PV) energy is the electrical energy that is derived from the direct conversion of light. By using specially made solar cells – or PV cells – light is absorbed and transformed directly and continuously into an electric current, which can be used to power an electric circuit. Technically speaking, a solar cell is a type of semiconductor that is sensitive to light. Figure 2.1 shows the basic operation of a typical PV cell.

Some semiconductors are sensitive to sunlight. When sunlight strikes them, the electrons (or particles of negative electric charge) within these semiconductors absorb the sunlight's energy, become excited and move around. Solar cells are thin, flat, 'wafers' of these light-sensitive semiconductors, with the two faces of the wafer separated by a 'junction'. Because of the special properties of each side of the semiconductor, the electrons tend to accumulate on one face, while the other is in deficit, creating a voltage between the two. When a wire connects the two faces, this voltage forces the electrons to flow through the wire from the side where they have accumulated to the side where there are few. This flow of electrons creates an electric current, which can operate a load, such as a light bulb or a radio.

The most commonly used photovoltaic cell material is *silicon* (see Figure 2.2). It is one of the most abundant elements on earth – ordinary sand is an oxide of silicon!

The first commercially available cells were *monocrystalline silicon*, in which all the silicon atoms are aligned in a highly organized crystal. These types of cells today have efficiencies of 14 to 17% and advanced small-area cells have reached 28% efficiency in the laboratory. Monocrystalline silicon cells are made of thin wafers (0.1–0.5 mm) of nearly pure silicon. This requirement for pure silicon necessitates an expensive manufacturing process.

In order to reduce costs, new manufacturing techniques were developed which resulted in *polycrystalline silicon* cells. While a monocrystalline cell is made of a single crystal, polycrystalline cells contain many crystals, each with the atoms aligned in a different direction, bound together. This cell type can be produced by a number of techniques that lend themselves to easier and faster production using less-pure silicon. The potential for cost reduction is deemed higher than for monocrystalline silicon. At 12 to 14%, commercial polycrystalline cells are only slightly less effi-

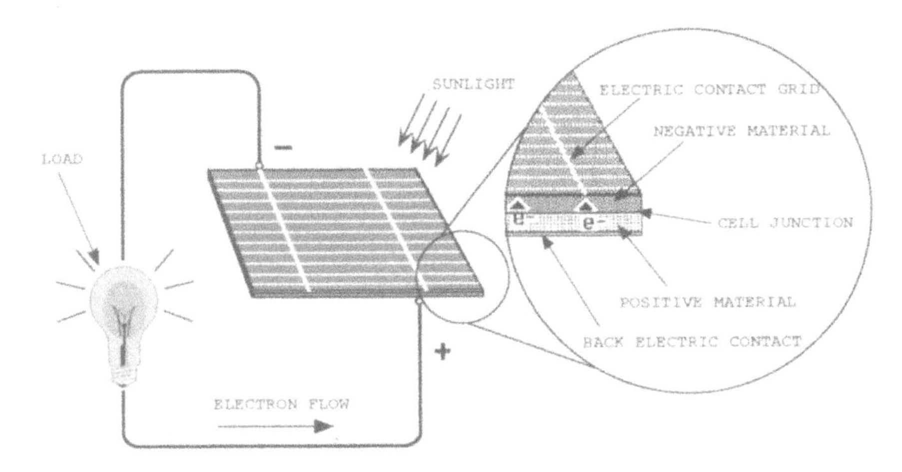

Figure 2.1. A photovoltaic cell

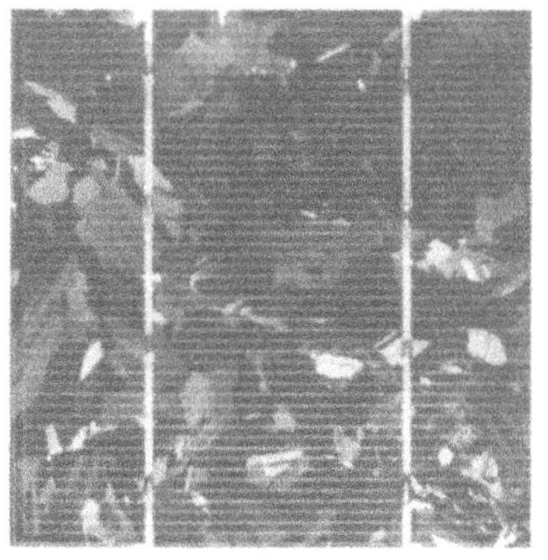

Figure 2.2. Monocrystalline, polycrystalline and flexible amorphous silicon cells (*photos:* Siemens Solar Industries, Solarex and United Solar Systems Corp., respectively)

cient than monocrystalline cells and are therefore widely used.

Development of *thin-film technologies* has permitted further cost reductions by reducing the amount of material needed to make a cell. *Amorphous silicon* modules require only a thin layer (typically 0.001 mm thick) of silicon and they can be mass produced automatically. Unlike monocrystalline and polycrystalline cells, amorphous cells have no crystal structure – atoms are arranged randomly within the cell. While efficiency is around 8% to 10% when new, instability at the molecular level lowers efficiency to about 3% to 6% after a few months exposure to sunlight. Because this low efficiency means that more cells are required to generate a given amount of power, amorphous silicon technology is rarely used in remote cold-climate areas where transportation is costly. New production techniques have been developed to manufacture 'multijunction' amorphous cells that contain two or three layers of semiconductor. These have higher stabilized efficiencies, ranging from 5% to 9%.

Other thin-film technologies have been developed but are not yet widely available. Two of the most promising are *cadmium telluride* and *copper–indium diselenide*. Cadmium telluride modules are beginning to appear on the market. In future, these may substantially reduce the cost of photovoltaics.

THE BEHAVIOUR OF PHOTOVOLTAIC CELLS: THE *I–V* CURVE

Like a chemical storage battery, a photovoltaic cell produces electricity as direct current (DC). Unlike a battery, which acts as a voltage source, a PV cell acts like an imperfect current source. For example, car batteries are typically 12 V. Regardless of how much current is required to turn the starter motor and operate the headlights, the voltage across the battery terminals stays close to 12 V. In contrast, at a given light level, a PV cell's voltage will vary while the current will stay relatively stable up to a certain voltage. When all the possible combinations of voltage and current are graphed for a certain light level, the result is the *I–V curve*, so named because it shows the characteristic relationship between current *I* and voltage *V*. An *I–V* curve for a typical crystalline silicon cell is shown in Figure 2.3; by convention, current is on the vertical axis and voltage on the horizontal axis.

The PV cell can be operated at any combination of current and voltage on its *I–V* curve, but in reality it operates at only one combination at a given time. This favoured combination is chosen not by the cell, but rather by the electric characteristics of the circuit that is connected to the cell. Consider a cell that is completely unconnected: no current is able to flow, so the current is 0 A (amperes). From the *I–V* curve, it can be seen that at 0 A the cell operates at a voltage of nearly 0.6 V. The voltage that occurs when the current is 0 is known as the *open-circuit voltage* (V_{OC}) and is the highest voltage that the cell can attain at a given light level and temperature (see Figure 2.4). On the other hand, if a wire is connected between the two sides of the cell, the wire will short-circuit the cell and force both sides to the same voltage. Thus the voltage across the cell will be 0 and the *I–V* curve will dictate a current of about 3.2 amps – the *short-circuit current* (I_{SC}), the maximum current attainable at this light level and temperature.

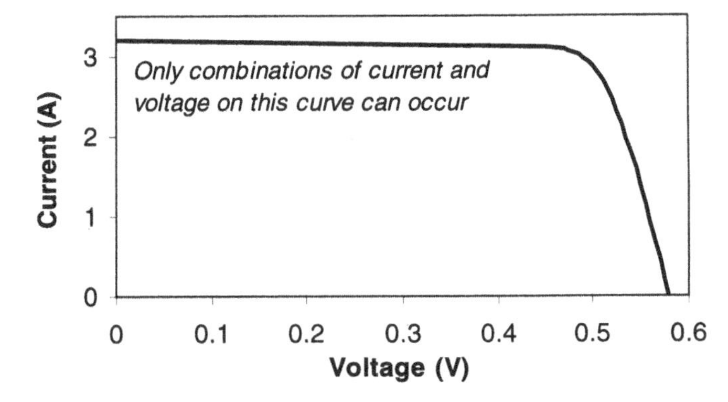

Figure 2.3. The *I–V* Curve for a typical silicon cell under standard test conditions

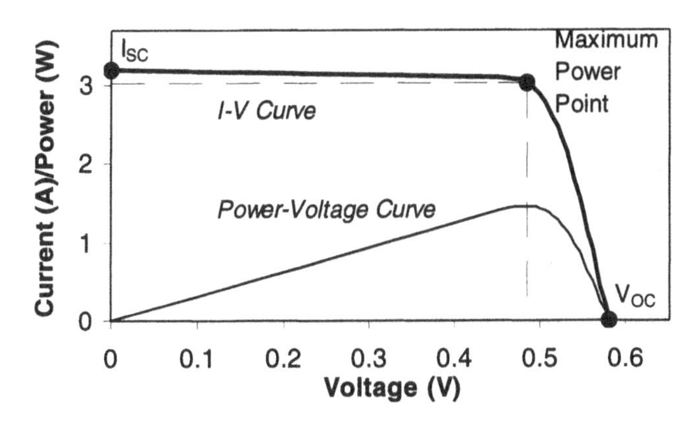

Figure 2.4. Important points on the *I–V* curve

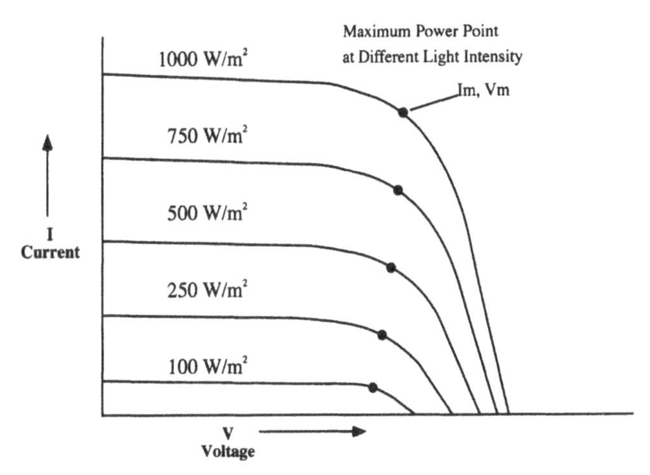

Figure 2.5. The effect of light intensity on the *I–V* curve

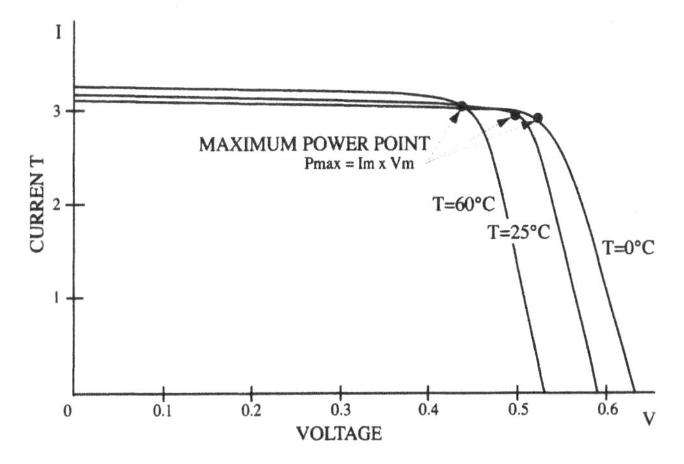

Figure 2.6. Influence of temperature on cell energy production; the maximum power output of the cell rises by 10% when the cell temperature drops from 25°C to 0°C

While current and voltage are at a maximum under short-circuit and open-circuit conditions respectively, the power (i.e. the product of current and power) at these points on the *I–V* curve is zero (see Figure 2.4). This is not very useful and real circuits are matched to the *I–V* curve so that they operate at a combination of current and voltage at which a reasonable amount of power is produced. The optimum point is the *maximum power point*, which for this cell occurs at about 3.0 A and 0.5 V, generating about 1.5 W (watts). The *efficiency* of a cell is the ratio of its power output when operating at its maximum power point to the power of the sunlight striking the cell.

The *I–V* curves shown in Figures 2.3 and 2.4 are for a particular cell under particular conditions. Other cells and other conditions will lead to different *I–V* curves. The important factors affecting the *I–V* curve are:

- *Light intensity.* Since PV cells convert the power of sunlight into electric power, it makes sense that their power output, and therefore their *I–V* curves, vary with the intensity of light. The shape of the *I–V* curve stays roughly the same, but at lower light levels the curve shifts downwards. Thus the short-circuit current is proportional to the light level and the open-circuit voltage changes only slightly (see Figure 2.5). In Figures 2.3 and 2.4, the cell is illuminated by 1000 W/m² of sunlight, or the approximate irradiance of the sun at noon on a bright day.
- *Size of the cell.* More sunlight strikes larger cells and, as a consequence, they generate more power. The short-circuit current is proportional to the size of the cell while the open-circuit voltage of the cell is not affected by it. Presently, most cells are about 100 cm² in area, and the *I–V* curves of Figure 2.3 and 2.4 are for such a cell.
- *Cell technology.* Cells made of different materials or by different processes differ in their *I–V* curves. The cell material determines the open-circuit voltage: for all crystalline silicon cells, for example, this will be around 0.6 V. The material also determines the cell's efficiency. Different materials will be more or less efficient in converting light into electricity.
- *Temperature.* Lower cell temperatures improve efficiency and raise open-circuit voltage – a good thing in cold climates! For example, the open-circuit voltage of a silicon cell will decrease by 2.2 mV for every degree Celsius rise in temperature (see Figure 2.6). This is discussed in more depth in Chapter 5.

Because the *I–V* curve of a photovoltaic cell is affected by so many factors, a set of 'standard test conditions' (STC) has been defined by international convention to evaluate its performance. At STC, the cell is illuminated by 1000 W/m²

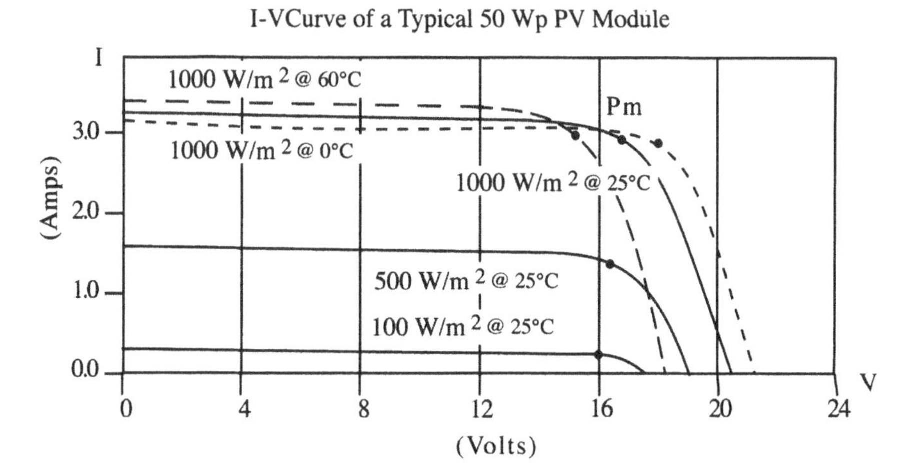

Figure 2.7. *I–V* curves for a typical 50 W_p module under various conditions

of sunlight and has a temperature of 25°C. Photovoltaic devices are rated in terms of their *peak watt output* (measured in W_p), which is their maximum power output at STC. Since the cell in Figures 2.3 and 2.4 is at STC, its peak watt output of 1.5 W_p can be read off the *I–V* curve.

COMPONENTS IN PHOTOVOLTAIC SYSTEMS

Modules and arrays

A single PV cell by itself is of little interest except in powering small electronic devices such as watches and calculators. In order to increase the operating voltage of a photovoltaic system, the cells are connected in series to form a PV *module*. Modules of 30 to 36 cells in series, configured specifically for the purpose of charging 12 V batteries, are common. As we have seen, the number and size of the cells will determine the maximum power that the module will produce at a set voltage. Figure 2.7 shows *I–V* curves for a typical 50 W_p silicon module. Note that the general shape of the curve and the short-circuit current are identical to those of the cells from which the module is built – only the voltage has changed.

The module is constructed in such a way as to protect the cells from the elements. The cells are typically encapsulated in a plastic material sandwiched between a sheet of tempered glass and a backing material. The backing is usually plastic, but sometimes aluminium or glass are used. Often the edge of the module is framed in aluminium, which provides strength and makes it easy to attach the module to a structure. Modules come in different shapes and sizes; they are the basic building blocks of a PV system.

As with the PV cells, a number of modules can be connected together in series to give a higher voltage or in parallel to obtain a higher current. A photovoltaic *array* consists of a number of electrically connected PV modules, associated cabling and interconnection components, a support structure and, when needed, special devices to protect the array against lightning, reverse flow of current and other electrical hazards (Figure 2.8).

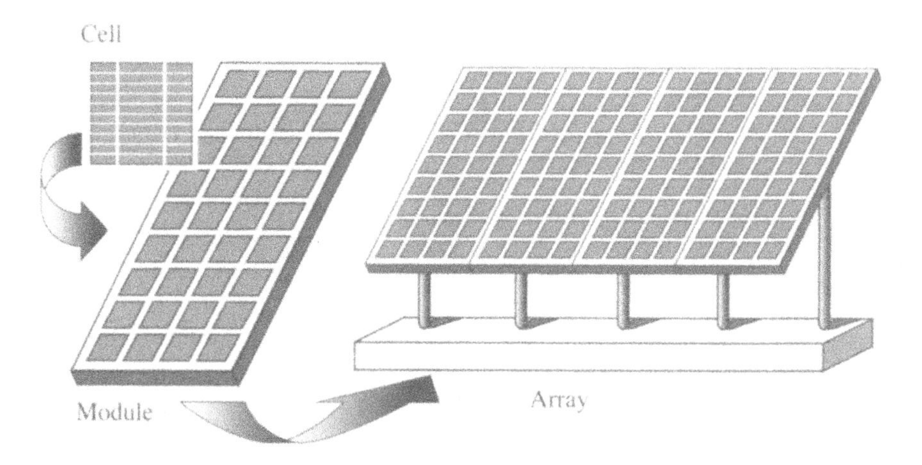

Figure 2.8. Conceptual relationship between the PV cell, module and array

Mounting structures allow the PV modules to be mechanically stable and withstand the environmental conditions they are facing. The PV array can be ground-mounted on a pole or on a frame. It can also be mounted directly on buildings using special roof or wall-mounting structures. The modules can even be integrated into the building's envelope and act as an environmental barrier – while producing electricity at the same time.

Mounting structures usually serve to position the modules for maximum light availability. Some structures allow the modules to remain facing directly towards the sun by automatically adjusting the tilt and/or orientation of the array. These special mounting structures are called *trackers* and are discussed in Chapters 3 and 5.

Batteries in PV systems

As we have seen, the output of a PV array depends on the available sunlight. In darkness, of course, the module produces no power. Such variation is permissible for a few applications, such as water pumping, but for most applications the loads require that electrical output be available on demand. Consequently, most PV systems use storage devices, the most common being *rechargeable batteries* (also known as 'secondary' batteries, in contrast to 'primary' batteries, which are not rechargeable). These are connected between the array and the load and store energy until it is needed.

A battery stores electric current in the form of chemical energy. When a battery is connected to a load, a chemical reaction occurs between the poles of the battery, which results in a flow of electrons through the battery to the load. In rechargeable batteries, this reaction is reversible: when electricity is 'applied' to the poles of the battery, the chemical reaction is reversed, returning the battery to its original chemical state. Just as modules are made from series-connected PV cells, batteries are made from series-connected electrochemical cells. Lead–acid cells are the most common type in photovoltaic systems; more expensive nickel–cadmium cells are used in some special circumstances.

Batteries fulfil three important functions in a photovoltaic system:

- *Autonomy*, by meeting the load requirements at all times, even at night or during overcast periods. During sunny days, the PV array generates enough power for the load and to charge the battery. During night or on cloudy days, the load is powered from the batteries, as shown in Figure 2.9.
- *Surge current capability*, by supplying, when necessary, currents higher than the PV array can deliver, especially to start motors and other equipment.
- *Voltage control*, by preventing large voltage fluctuations that could damage the load. When they are connected, a battery will force a PV array to operate at the battery's voltage; since the battery voltage remains relatively constant, this protects the load from wide voltage fluctuations.

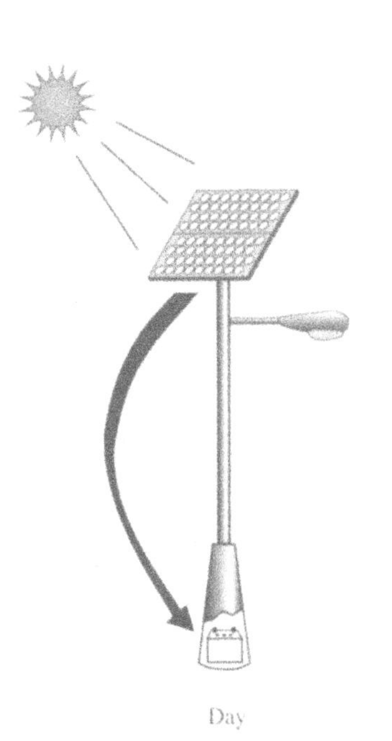

Figure 2.9. Batteries store excess PV energy during the day and power the load at night

Power-conditioning equipment

Usually the electrical output of an array is modified or regulated in some way to ensure that it meets the requirements of the rest of the system. The devices which do this are *power-conditioning devices*:

- *Battery charge regulators* are the most common type of power conditioners. Their primary function is to cut off charging current from the array when the battery reaches a certain voltage threshold, indicating that it is fully charged. Other functions of a regulator may include disconnecting the load from the battery when the latter is fully discharged and periodic battery 'equalization', in which a high voltage charge ensures that all cells in the battery are uniformly charged.
- *Inverters* are required in systems which supply power to AC loads. They convert the DC output of the array and/ or the battery to standard AC power similar to that supplied by the utilities.
- *Power-point conditioners* are devices which can vary the output voltage and current of the PV array so that it more closely matches the requirements of the load.

PHOTOVOLTAIC SYSTEMS

A photovoltaic system consists of the complete set of equipment used to convert light into a useful form of electricity. This includes the PV array, the conditioning equipment, the cabling, protective devices and, when needed, the storage system and an auxiliary power supply.

Photovoltaic systems can generally be categorized as either *autonomous*, which are not connected to an electric grid, or *grid-connected* (*grid-tied*), which are. Autonomous systems, also known as '*stand-alone*' systems, can be further subdivided according to whether they have battery storage and auxiliary power supplies (Figure 2.10). In all, there are four types of PV systems:

- *Stand-alone PV system without battery.* This system powers a load which does not require an electric storage system. This may be the case when the energy production from the PV system is adequate even in subdued lighting (e.g. calculators), when the operating time of the load is not critical or is matched to the availability of sunshine (e.g., ventilation fans), or when storage is integral to the system (e.g. water pumping, in which a water tank or reservoir usually provides storage).
- *Stand-alone system with battery.* This is the most common type of photovoltaic system. The PV array powers the load and charges a battery during the day. When little or no sunshine is available (for example, during the night), the battery powers the load.
- *Stand-alone PV-hybrid system.* In this configuration, an auxiliary power source augments the energy production of the array. During those periods when the average load is larger than the average output of the PV array, the auxiliary power source makes up the difference; a battery

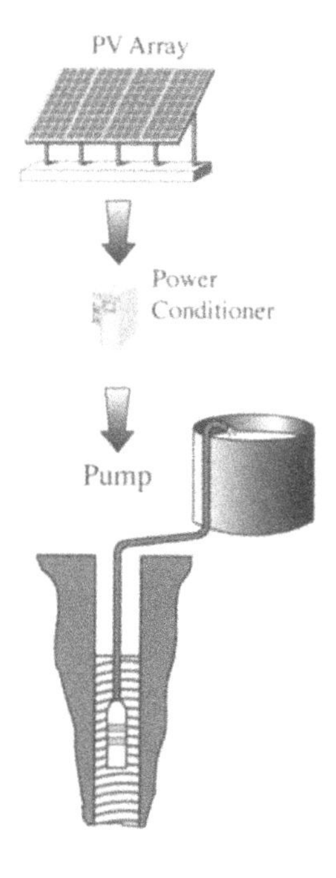

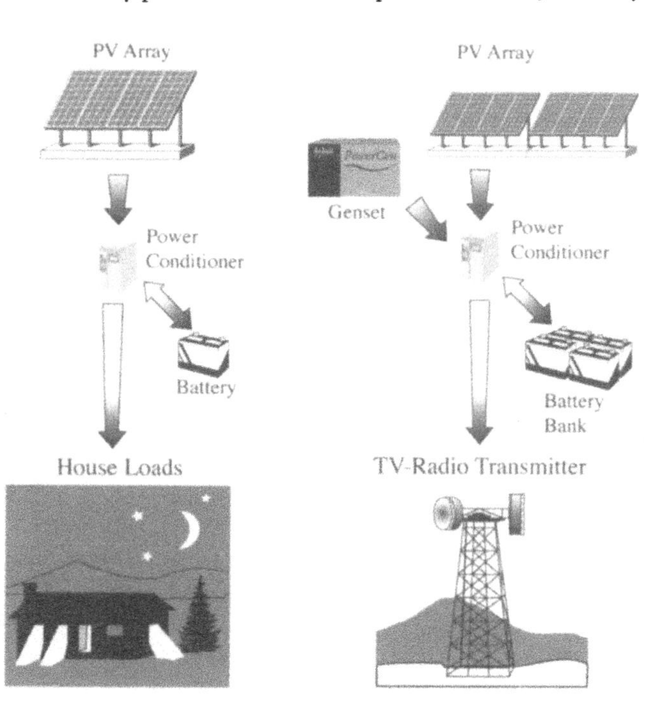

Figure 2.10. Types of autonomous systems: PV system without batteries, PV system with batteries and PV–hybrid system

usually buffers short-term variations in the availability of sunshine. The most common auxiliary power source is a generator powered by a diesel, gas or gasoline engine, called a *genset*. When the energy stored in the battery has been used up, the genset starts, powers the load and recharges the battery. The constant availability of the genset complements the variable input of the photovoltaic system.

- *Grid-connected PV systems.* A PV array can be connected to a grid by way of an inverter; the electricity generated by the array is then used by grid-connected loads and is indistinguishable from electricity generated by other sources (Figure 2.11). In this way, other electric generators on the grid need not generate so much electricity; if those other generators are powered by fossil fuels, this results in fuel savings and reduced pollution. The grid can be a major one – the North American power grid, for example – or a small isolated grid, such as those found in remote communities and at remote mines. The PV contribution to the grid can come from a single large installation – a '*centralized*' plant – and/or from numerous smaller arrays scattered around the grid in a '*distributed*' fashion. For example, some grid-tied buildings have PV arrays on their roofs: when sunshine is available, the array generates power for the building and feeds any excess onto the grid. When loads are large or the sun is not shining, the grid powers the building; a meter keeps track of the net use and supply of grid electricity. This type of system reduces losses inherent in the transmission and distribution of electricity over long distances and large areas and is especially well suited when peak loads usually occur during sunny periods – such as in areas where air conditioning is common.

ADVANTAGES OF PHOTOVOLTAIC ENERGY

Photovoltaic energy has a number of advantages compared to competing power sources:

- *High reliability.* PV modules have no moving parts to wear out and they last more than 25 years. Two manufacturers presently guarantee that module output will not degrade by more than 20% over 20 years. Photovoltaic systems have been successfully used in unmanned, environmentally hostile sites for numerous applications around the world.
- *Low operation and maintenance costs.* While photovoltaics are expensive to purchase, they do not have the high operation and maintenance costs of conventional power sources. These costs can be very significant in remote areas, where transportation is expensive and time-consuming: maintenance personnel and fuel may have to be flown in by plane or helicopter, making it more costly to repair equipment than it is to purchase it.
- *Low risk.* PV systems are capital intensive but have low operating and maintenance costs. Once installed, they are

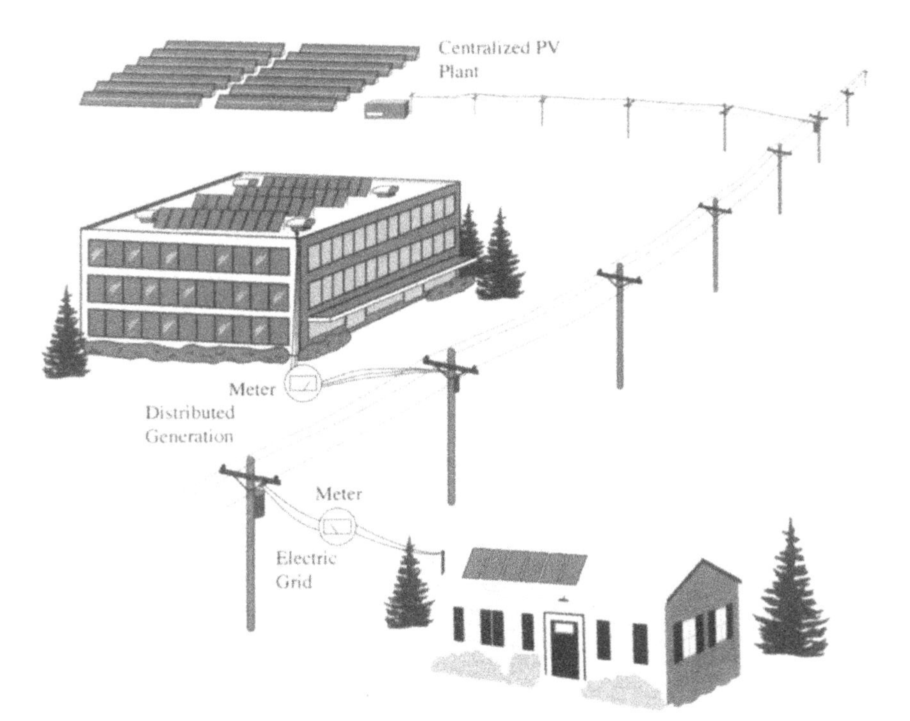

Figure 2.11. Grid-connected PV systems

not subject to exterior costs and their operating costs as well as their energy output are relatively stable and predictable. The financial risks of PV systems are lower than those of conventional systems: the cost of the project will not escalate as a result of changes in fuel prices or operation and maintenance costs.

- *Environmentally friendly.* PVs are one of the most benign ways to generate electricity. A PV power system needs no fuel and has low maintenance requirements. PVs do not produce fumes, noise or radiation and do not need cooling water. All that is really required is space to install the PV arrays. This space does not have to be prime land and PV modules can be integrated into existing structures such as roofs.

  While the production of PV devices involves industrial processes, the environmental impact of these processes is limited and controllable. PV manufacturing plants are strictly regulated and must comply with standard health and safety requirements. Although some thin-film technologies contain hazardous substances, these are 'fixed' in the cell, are very stable and pose little threat to the environment. Furthermore, such technologies can be recycled at the end of their useful life.
- *Versatility.* Since the 'fuel' source is omnipresent, centralized power production is unnecessary. It is usually cheaper and easier to install smaller PV power systems where their power is needed, thus reducing losses of electricity during transmission and distribution.
- *Modularity.* PVs can be used in applications with very small energy requirements (calculators, watches) as well as those requiring large amounts of energy (grid power). A number of the same building blocks, the PV module, are interconnected to obtain the required power and voltage. This also allows for build-up; modules can be added to existing plants in order to increase power output. Construction of a PV generating plant is fast, taking less than a year to build MW plants. Little specialized knowledge is needed to build a large PV plant.
- *Rapid progress.* A substantial international R&D community is actively pursuing PV development in a number of areas to reduce the cost of PV and facilitate its use. New advances are reported regularly, indicating that performance and cost limits have not been reached. R&D activity in other areas, such as semiconductor materials for electronics or new chemistry for batteries, benefits PV technology directly.

FURTHER READING

Komp RJ (1984). *Practical Photovoltaics: Electricity from Solar Cells.* 2nd edn. Ann Arbor, MI: Aatec Publications.

Lasnier F, Gan Ang T (1990). *Photovoltaic Engineering Handbook.* Bristol, UK: IOP Publishing.

Maycock PD, Tirewalt EN (1985). *A Guide to the Photovoltaic Revolution.* Andover, MA: BrickHouse Publishing Co.

Zweibel K (1990). *Harnessing Solar Power: The Photovoltaics Challenge.* New York, NY: Plenum Press.

3 The solar resource in cold climates

Bengt Perers
(Vattenfall Utveckling AB)

Sunshine, the fuel of every PV system, is different at the high altitudes and high latitudes where cold climates occur. At high latitudes, days are long during the summer and short during the winter, and the sun stays low in the sky. As a result, the seasonal variation in the solar resource is much more pronounced than at more equatorial locations – a lot of energy is available during summer but very little is available during winter. At high altitudes, the thinner atmosphere leads to slightly more intense sunshine. These effects, as well as others, must be taken into consideration when designing PV systems for cold climates.

INTRODUCTION

The amount of sunshine falling on a photovoltaic array depends not only on the strength of the sunlight, but on the orientation of the array, especially in relation to the position of the sun. Before discussing the solar resource, therefore, some basic terms should be defined.

The orientation of the array is described by two angles: the *tilt angle* is the angle formed by the array and the horizontal; the *surface azimuth angle* specifies where the array faces (more exactly, it is the deviation from the local meridian of the projection on a horizontal plane of the normal to the surface), with due south being 0°, due west +90° and due east –90° (see Figure 3.1).

As the sun moves across the sky, light rays strike the array from different directions. This is measured by the *angle of incidence*, which is the angle formed between the light ray and a line normal to the array (see Figure 3.2). When the sun's rays strike the surface of the array squarely, the angle of incidence is 0°; when they shine parallel to the surface of the array, the angle is 90°. To maximize the solar energy collected by an array, the angle of incidence should be kept as near to 0° as possible, as shown in Figure 3.2. Some arrays physically track the sun in order to do this; however, most arrays are fixed in place.

Sunlight is scattered in the atmosphere and reflected by the ground; the light striking a photovoltaic array comes not just from the direction of the sun, but from other directions as well (see Figure 3.3). The *total solar radiation* is the total amount of radiation incident on the array. The *beam radiation* (also referred to as direct radiation) is the radiation coming from the direction of the sun; it is sunlight that has not been scattered by particles in the atmosphere. *Diffuse radiation* is the opposite: it comprises scattered radiation from all directions except the disc of the sun. *Ground-reflected radiation* arrives at a surface from the direction of

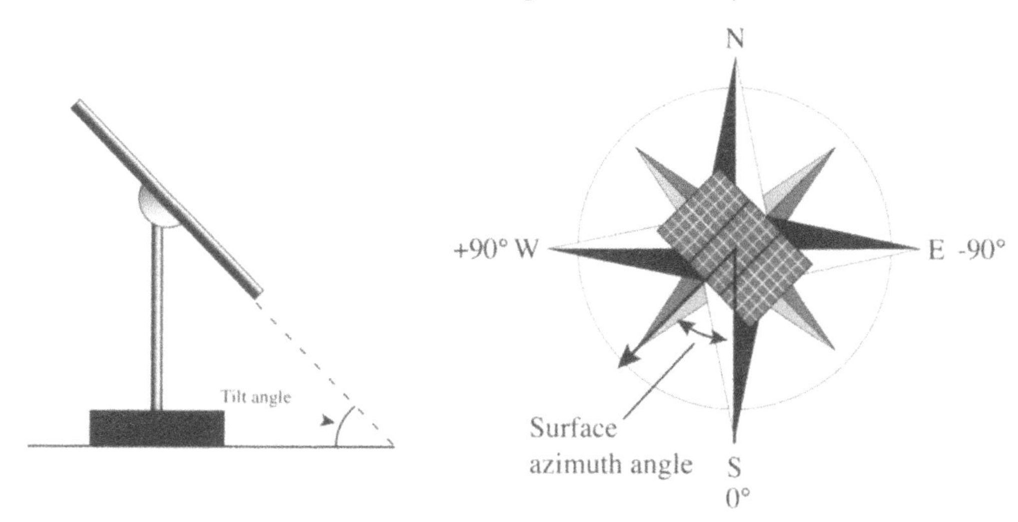

Figure 3.1. Tilt angle and surface azimuth angle

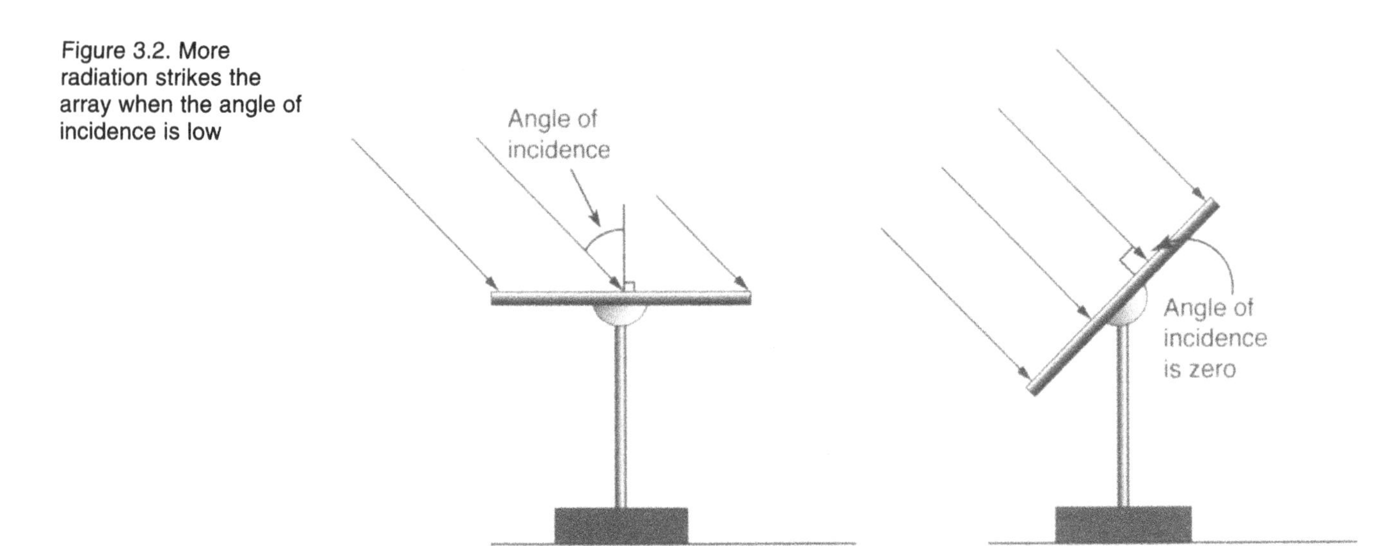

Figure 3.2. More radiation strikes the array when the angle of incidence is low

the ground – it is the sunlight that has 'bounced' off the ground. The total solar radiation incident on a horizontal surface is often called the '**global**' radiation on the horizontal (plane).

Irradiance refers to the *power* of the sunlight striking a surface. It is watts per square metre of sur ıny day, a surface angled towa 000 W/m^2. Under certain cond ceed this level for a short time

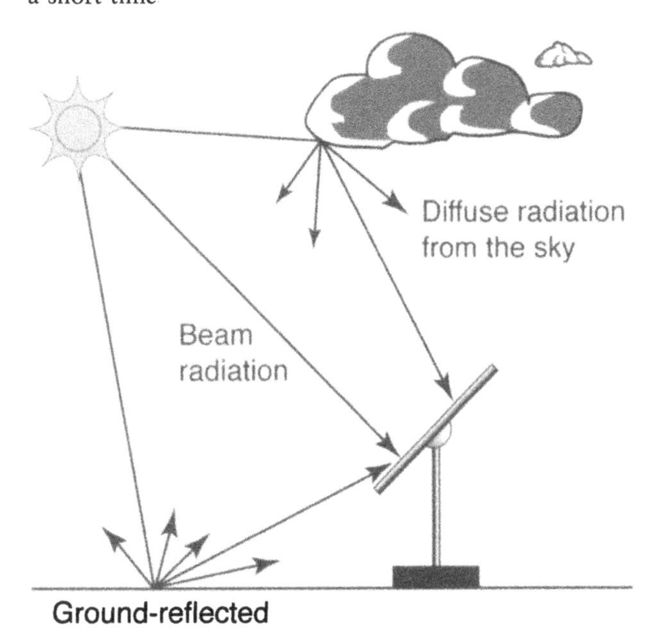

Figure 3.3. Beam and diffuse radiation

Insolation is the *energy* contained in the sunlight striking a unit area of surface *over a specified period of time*; it is, therefore, the measure of *solar irradiation* and is usually measured in units of joules per square metre (J/m^2) or watt-hours per square metre (Wh/m^2); one Wh/m^2 is equal to 3600 J/m^2. Another convenient unit for insolation is the *peak sun hour*, which is equal to the solar energy striking a one square metre of surface in one hour when the irradiance is 1000 W/m^2. For example, during the course of an average June day in Stockholm, a total of about 6.5 kWh, 6.5 peak sun hours or 23.5 MJ of solar energy strikes a one square metre of horizontal surface. Note that, because watt-hours and joules are small units, kilowatthours (kWh) and megajoules (MJ) are often employed. One thousand Wh equal one kWh and one million J equal one MJ.

VARIATION IN THE SOLAR ENERGY RESOURCE

The amount of sunlight falling on a given photovoltaic array varies greatly from one point in time to another; broadly, there are two sources of variation:

- *Celestial movement.* Movement of the earth with respect to the sun gives rise to the seasons and the cycle of day and night.
- *Atmospheric conditions.* Gases and particles in the atmosphere scatter and absorb sunshine.

Day, night and the seasons

Day and night are caused by the rotation of the earth. The earth spins once a day on an axis that runs through the North and South Poles. Viewed from the North Pole, the earth spins counter clockwise, giving the impression that the sun rises in

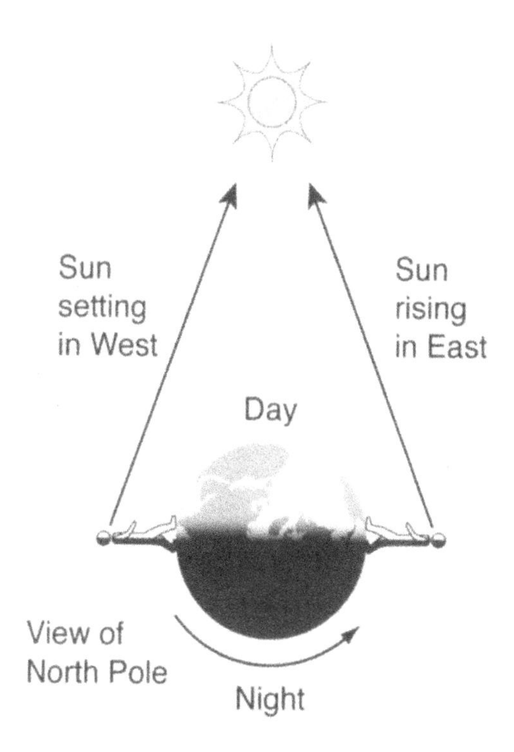

Figure 3.4. The rotation of the earth causes day and night

the east, arcs over the sky and sets in the west (see Figure 3.4).

We set our watches more or less according to the sun's position in the sky. For example, *solar noon* is the time of the day when the sun's shadow runs perfectly north–south. Noon on a watch is roughly equal to solar noon, though differences arise as a result of the use of standard time. In addition, during summer some regions use daylight-saving time, moving the time on a watch a further hour ahead of the solar time.

Two things cause the seasons. First, the earth revolves around the sun with a period of 365 days. Second, the axis of the earth's rotation is tilted at 23.5° to the plane of the earth's orbit. Were the axis not tilted, day and night would always be 12 hours each. Since it is tilted, the axis points in different directions at different points in the earth's orbit. During summer in the northern hemisphere, the North Pole points towards the sun and the northern hemisphere enjoys long days. During winter, the North Pole points away from the sun and the days in the southern hemisphere are long. During spring and fall, the North Pole points between these two extremes (Figure 3.5).

This effect is exaggerated at higher latitudes (Figure 3.6). At the equator, day and night each last 12 hours, regardless of the season. At the poles, the sun circles the sky for 186 days without setting ('polar day') before it finally dips below the horizon, not to rise for 179 days ('polar night'). The length of day varies at latitudes between these extremes; for example, at the Arctic circle (66.5° N) and Antarctic circle (66.5° S), there are only 24 hours of polar day and 24 hours of polar night per year.

Like the length of day, the sun's trajectory across the sky changes from season to season, with the change being more pronounced at higher latitudes. At the equator, the sun always rises more or less in the direction of due east, passes

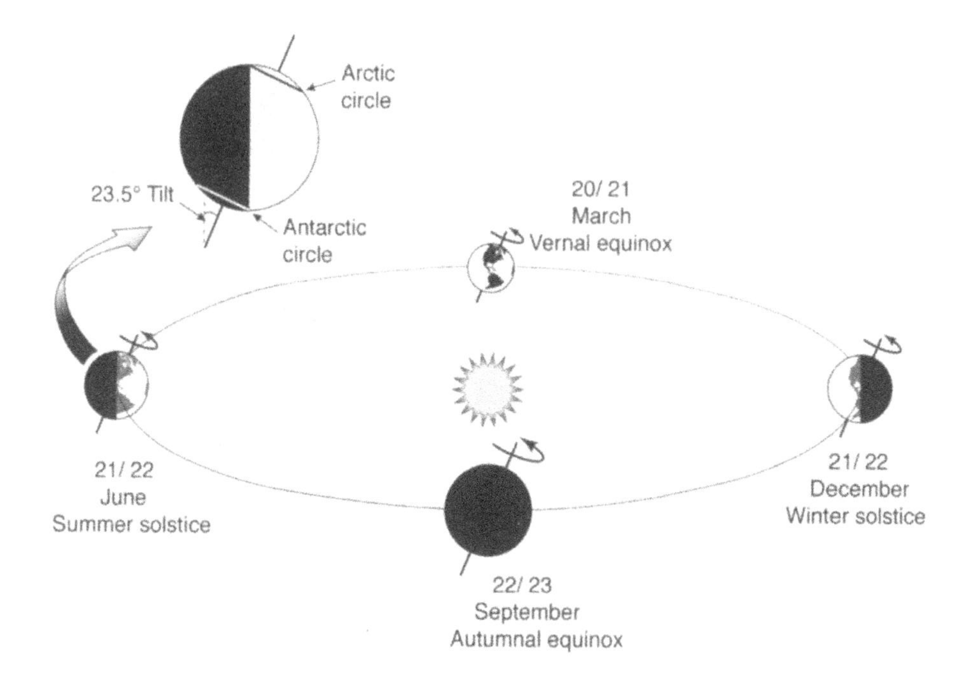

Figure 3.5. The seasons, the equinoxes and the solstices (adapted from Iqbal[1])

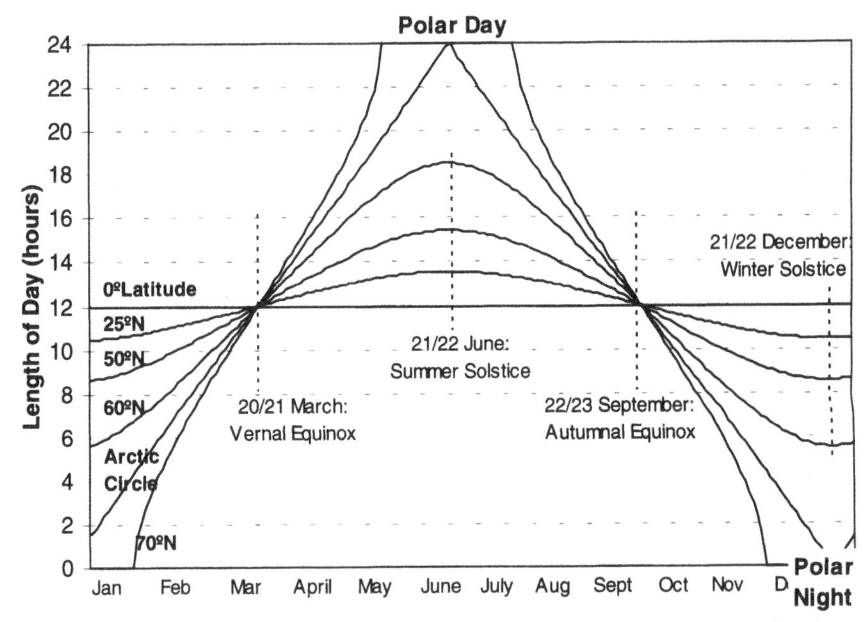

Figure 3.6. Length of day at different latitudes in the northern hemisphere; for the southern hemisphere, the curves are inverted

nearly overhead at noon and sets in a direction close to due west. Off the equator, however, the sun will spend half the year rising and setting in the southern sky and half the year rising and setting in the northern sky. For example, during summer in the northern hemisphere, the sun rises in the north east and sets in the north west. This means that the sun will rise and set *behind* a south-facing fixed array. In fact, whenever daylight lasts longer than 12 hours, the sun will be behind a fixed array for the portion of the day in excess of 12 hours. The shorter the day, the closer to due south the sun rises and sets (in the northern hemisphere). Furthermore, at higher latitudes, the sun's path between sunrise and sunset will stay closer to the horizon: at noon, the sun will be lower in the sky (see Figure 3.7), affecting the angle of incidence.

Because the sun can be low in the sky, especially during winter at high latitudes, one must not confuse radiation on the horizontal plane with radiation falling on the array. For example, at Stockholm, Sweden (59°N), the insolation on a horizontal surface is 26 times larger in June than December, but for an appropriately tilted array it will be only five to eight times larger.

Atmospheric effects

Outside the earth's atmosphere, a surface facing the sun receives approximately 1370 W/m^2 of solar radiation. As this solar radiation passes through the atmosphere, it encounters air molecules, dust, water droplets and ice crystals,

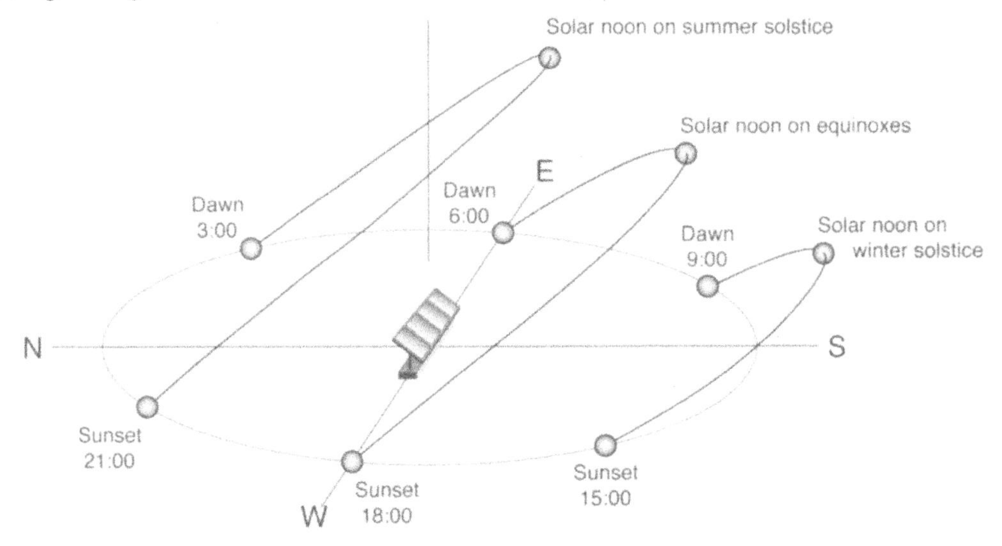

Figure 3.7. Apparent position of the sun at various times of the day and year at Stockholm, Sweden

which scatter and absorb some of the sunlight, reducing the amount reaching the earth's surface. The extent to which the radiation is reduced depends on the clearness of the atmosphere and the thickness of the layer of atmosphere that the sunlight must pass through.

Under clear skies, a surface angled towards the sun will receive around 1000 W/m^2 of sunlight at noon; under heavily overcast skies, the radiation may be only 5% of this. Clouds and haze not only diminish the amount of sunlight reaching the ground, but they cause diffusion. On a clear day, beam radiation will exceed diffuse radiation by a factor of about ten; on a cloudy day, nearly all radiation will be diffuse.

The relative positions of the sun and the earth – and consequently the length of day and location of the sun in the sky – can be calculated to great precision using simple formulae; the presence of clouds cannot, as errors in the weather forecast frequently demonstrate. This means that predicting the amount of sunshine that will strike a surface is not a simple matter of geometry, but rather involves using data from past years to estimate future sunshine. In photovoltaic systems without a back-up power system, a key problem is determining the minimum solar energy that will fall on the array over a given period. Once this has been estimated, the array can be sized so that it will provide sufficient electricity. Unfortunately, there is always a finite probability that less sunshine will fall on the array than would be expected on the basis of historical data. This possibility can never be eliminated; it stems from the natural, unpredictable variation in the weather.

Clearness may vary with the season. In many cold climate areas, winter is typically less clear than summer. In coastal climates, the average clearness during winter months can be half that of summer months. At most sites, however, the seasonal variation in *clearness* does not exceed 30%. Note that the variation in *insolation* will be much larger than this at mid- and high-latitude sites.

The altitude and the sun's position in the sky determines how much atmosphere sunlight must pass through before reaching the ground. An '*air mass*' scale is used to express the thickness of this atmospheric layer. Outside the earth's atmosphere, sunlight passes through no air, so the air mass is zero. When the sun is directly overhead, it passes through an air mass of one in reaching sea level. As the sun moves away from the point directly overhead, it passes through more and more atmosphere. For instance, when the sun is 30° above the horizon, it passes through twice as much air as when it is directly overhead, and therefore the air mass is two.

Even when there are no clouds, air attenuates solar radiation. Thus, thicker layers of atmosphere let less sunlight through. The sun stays close to the horizon in mid- and high-latitude systems, especially during winter; as a result, sunshine will be reduced somewhat.

At high altitudes, there is less atmosphere to block sunshine, so the maximum intensity of solar radiation is higher;[2] of course, it can never exceed the 1367 W/m^2 available outside the earth's atmosphere. Weather tends to be more variable in mountainous regions and a higher maximum intensity may not necessarily reflect higher average levels of sunshine.

THE SOLAR SPECTRUM

As demonstrated by the rainbow, sunlight consists of light of different colours. Colour is determined by the wavelength of the light. Outside the earth's atmosphere, 97% of the sun's energy is contained in the range of wavelengths from 0.3 to 3.0 microns; about half of this is visible light, a little less than half is infrared light and the remaining 5% is ultraviolet light. The distribution of the energy in sunlight as a function of wavelength is referred to as the *solar spectrum* (see Figure 3.8).

As sunlight passes through the earth's atmosphere, absorption of certain wavelengths modifies the spectrum. The extent of this change depends on the clearness of the atmosphere and the thickness of the layer of air that the sunlight is passing through, that is, the air mass. When the sun is low in the sky, as at sunrise and sunset and during winter, the air mass is high and the spectrum becomes relatively more red. When cloudy, it is made more blue. Furthermore, when light is reflected off a coloured surface, such as the ground, certain wavelengths are absorbed. This will affect the spectrum of light containing much ground-reflected radiation.

Photovoltaic cells are sensitive to only a certain range of

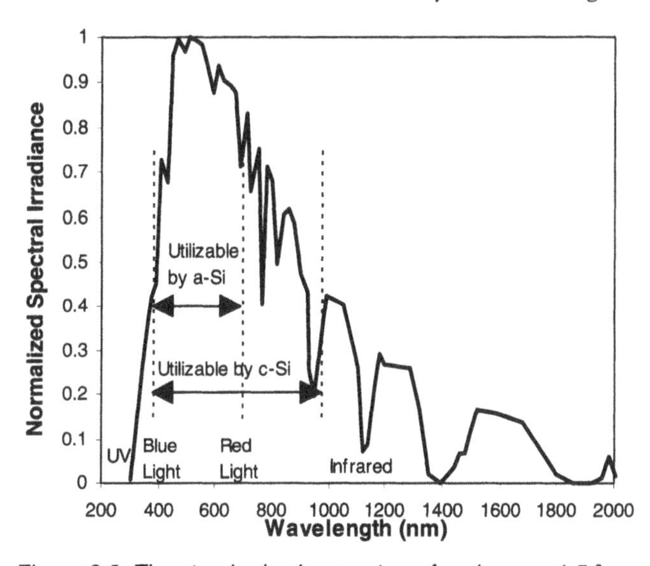

Figure 3.8. The standard solar spectrum for air mass 1.5;[3] the sensitivity limits of amorphous silicon (a-Si) and crystalline silicon (c-Si) are shown for comparison

wavelengths and thus variation in the solar spectrum can affect cell efficiency. By convention, efficiency is measured with the cell illuminated by a spectrum corresponding to that which occurs with an air mass of 1.5 and relatively clear skies. In real life, the spectrum will differ from this reference spectrum as a result of variations in air mass, clearness and composition of the atmosphere.

Fortunately, the effect of spectrum on efficiency turns out to be quite small. Most PV cells use crystalline silicon and are responsive to a relatively wide band of wavelengths; as long as the proportion of the sunlight's total energy that falls within this band does not change much, the efficiency will stay relatively constant. Dramatically shifted spectrums are rare and, when they do occur, such as at sunrise and sunset, they generally last only a short period of time and do not significantly affect long-term photovoltaic system performance.

While variation in spectrum does not significantly affect the long-term performance of monocrystalline cells, it can be important to their instantaneous efficiency. This can reduce the accuracy of photovoltaic devices, such as reference cells and photodiodes, used to measure the intensity of solar radiation. These sensors will be less accurate in cold climates, where the spectrum will be influenced by high air mass and ground-reflected radiation (see Chapter 14).

Amorphous silicon cells are less responsive to red light than are crystalline silicon cells (Figure 3.8). As a consequence, they are more seriously affected by the redder spectrum associated with high air masses. This will reduce efficiency when the sun is low in the sky, i.e. during winter and at high latitudes. For example, a 10% reduction in efficiency, attributed to spectral mismatch, has been observed during December in Austria.[4]

Not all parts of the spectrum will be equally affected by a change in altitude. Unfortunately, the affects of atmosphere on the spectrum are very dependent on the composition of the atmosphere and few generalizations can be made. One salient characteristic of solar radiation at high altitudes, however, is that it contains much more ultraviolet (UV) radiation than at sea level. Since PV panels are not sensitive to ultraviolet light, this will not increase their production. Strong UV rays will, however, accelerate ageing of plastics, particularly plastic insulation on cables and the clear plastic that encapsulates the PV cells within a module. One criterion for the selection of modules and cables is, therefore, their resistance to UV degradation.

GROUND REFLECTION

Beam and diffuse radiation from the sky can be reflected onto the array by the ground and objects near the array, increasing the output of the array (see Figure 3.3). The *albedo* is the ratio of sunlight reflected by a surface to the sunlight incident on the surface. Light-coloured surfaces like snow, water, white gravel or white walls reflect sunlight better than dark surfaces and thus have a higher albedo. Table 3.1 gives the albedo for a variety of common surfaces. A range of values is given because the albedo will depend on the nature of the surface (for example, grass has different albedos in spring and in autumn), the mix of diffuse and direct sunlight, the angle of incidence and the wavelengths under consideration.

Table 3.1. Typical albedo of some common materials[5]

Surface	Typical albedo
Fresh snow	0.7 to 0.9
Aged snow	0.6 to 0.8
Light-coloured paint	0.5 to 0.7
Ice	0.4 to 0.5
Melting snow	0.3 to 0.4
Grass	0.2 to 0.3
Sand	0.1 to 0.4
Soil	0.1 to 0.4
Concrete	0.1 to 0.3
Asphalt	0.1 to 0.2
Green forest	< 0.1

ENHANCEMENT OF RADIATION BY OPTIMUM LOCATION, TILT AND DIRECTION

The location, tilt and orientation of the array are normally the only factors that the user can influence in order to improve the performance of a photovoltaic array. When optimally oriented, a module will receive more light and produce more energy. Unlike some solar thermal collectors, a PV module's output is almost linearly dependent on the incident radiation, so even small enhancements in incident radiation can be utilized.

Optimum location: avoid shading

Shading is one of the most important factors to take into account when designing a system and calculating the available solar radiation. *The photovoltaic array must be evenly illuminated – shading of the PV array decreases its output greatly.* Shading is especially problematic in cold climates: at high latitudes the sun is lower in the sky and shadows are longer. It is easy to forget, when installing a PV array during the summer, that shadows will be longer during winter, just when solar energy is at a minimum. To minimize shading, the array should be located as high above the ground and as far away from trees and other shading objects as possible. Arrays consisting of a field of sub-arrays arranged in parallel

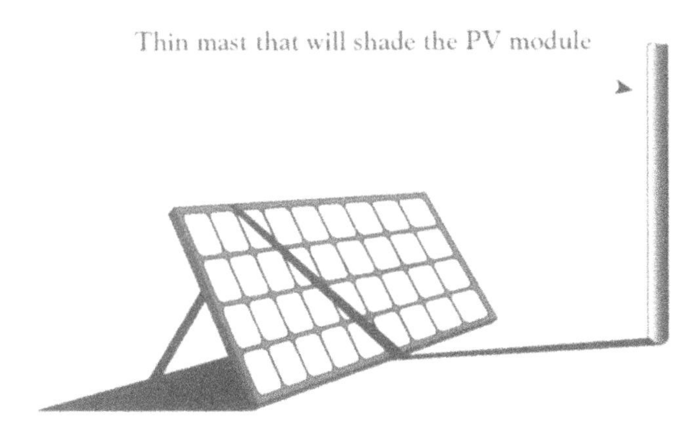

Figure 3.9. Minor shadows reduce PV module output dramatically because the solar cells are serially connected

rows must be spaced such that the front rows will not shade the rows behind them.

Merely shading one cell in a module will reduce the output of the module very significantly. In an array consisting of series-connected cells and modules, all the current must flow through each cell. Shaded cells will resist this flow of current. It is imperative, therefore, that no objects, however thin or small, shade the array (see Figure 3.9). This includes buildings, trees, poles, masts, guy wires, chimneys, vents and TV antennas. Note that in areas prone to icing, slender objects such as guy wires may present a much larger profile when covered in ice.

At high altitudes a PV system will often be surrounded by mountains, which will shorten the day and block direct sunlight. Similarly, when PV systems are installed along pipelines or transmission lines that cut through a forest, the trees on either side will tend to limit solar energy collection in the morning and afternoon ('alley shading').

When shadows cannot be avoided, the array must be enlarged so that it still produces the required amount of electricity. Software and nomograms for calculating when shading will occur and how much this will affect the PV system should be used (see Chapter 11).

Optimum direction

For fixed arrays, the optimum surface azimuth angle of the array is usually due south in the northern hemisphere and due north in the southern hemisphere, i.e. the array should face the equator. There are three cases when this may not be optimal, however:

- When there is a tree, building, mountain or other object that will cause shading in the morning, the array should be oriented towards the west, to favour afternoon sun. Analogously, if afternoon shading is expected, orient the array towards the east. Of course, it is best to locate the array so as to avoid the shading altogether, as mentioned above.
- In an autonomous system with a load occurring predominantly in the morning, it may be advantageous to orient the array towards the east, to favour morning sun. Orient towards the west for afternoon loads.
- In those coastal and mountain areas where the sky is typically clear in the morning but cloudy in the afternoon, orient the array towards the east, to favour morning sun. If it is typically clearer during the afternoon than the morning, orient the array towards the west.

For annual output, the surface azimuth angle is normally not critical within ±45° of due south (northern hemisphere) or due north (southern hemisphere) – losses will typically be less than 7%. Azimuth angle is even less critical for summer-only systems. On the other hand, for systems where the winter output is paramount the short days will give a smaller tolerance of about ±15° (for losses less than 4% at latitudes of 45° to 60°) to ±25° (for losses less than 6%).

Optimum tilt angle

As mentioned earlier, to maximize the collection of solar energy, the array must face directly towards the sun, i.e. the angle of incidence must be kept at 0°. When an array is fixed in place, it is impossible to do this, because the sun is moving across the sky. Instead, the array is usually oriented such that at midday, when the sun is strongest, the angle of incidence is near 0°. This is achieved when the array is facing the equator (see previous section) and when the tilt angle is optimized according to the latitude of the site.

To maximize *winter production*, set the array tilt to the latitude of the site plus approximately 15° to 20°: this will orient the array directly towards the sun at noon during the winter. Note that at sites near or above the Arctic or Antarctic Circle, tilt angles lower than latitude plus 15° may be preferred. At these sites, the sun will not rise in the middle of winter, so no energy is produced, regardless of the tilt angle. Selecting a lower tilt angle – equal to the latitude, say – cannot detrimentally affect winter production but will improve performance during the rest of the year.

If energy production during winter is important, the tilt angle should never be so low as to inhibit snow from sliding off the array. Snow will usually slide off panels that are tilted at 45° or higher; in areas with especially low temperatures or especially high snowfall, higher tilt angles may be required.

At sites within 30° of the equator, *annual energy production* is maximized when the array tilt is set roughly equal to the latitude. At higher latitudes, however, tilt angles somewhat lower than the latitude are better. Low tilt angles favour summer production and high tilt angles favour winter production; at high latitudes, so much more energy is available during summer that tilt angles should favour summer production whenever maximizing annual energy production is the goal.

For *maximum production from May through July*, tilt angles significantly lower than the latitude are best. These low tilt angles make sense since the sun sweeps across the sky during summer and a more horizontal module will capture more energy when the sun is to the east and west than a module that is more vertical. For the spring and autumn, module tilt angles roughly equal to the latitude are reasonable. Optimal panel tilt angles for various seasons and latitudes are illustrated in Figure 3.10 and Table 3.2. Note in Figure 3.10 that the panel tilt angle is not critical within about ±10°.

At sites that are very cloudy and where the albedo is low, slightly lower winter tilt angles may be better. High tilt angles maximize the utilization of beam radiation but, since the array 'sees' less of the sky, decrease utilization of diffuse radiation from the sky. If the albedo is high, e.g. snow, this will not reduce the total energy collected, since the ground

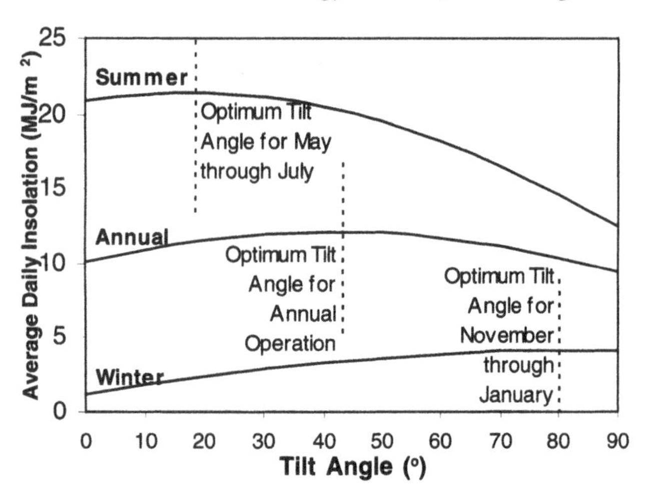

Figure 3.10. Insolation on the plane of the array for Stockholm, Sweden (59°N) as a function of array tilt angle

Table 3.2. Examples of optimal array tilt angles for different sites and different seasons; snow and dust accumulation is ignored

Site	Latitude	Optimal for December	Optimal for annual	Optimal for May to July
Miami, USA	26°N	45°	25°	5°
Montreal, Canada	45°N	65°	35°	15°
Stockholm, Sweden	59°N	80°	45°	20°
Resolute, Canada	75°N	N/A	55°	40°

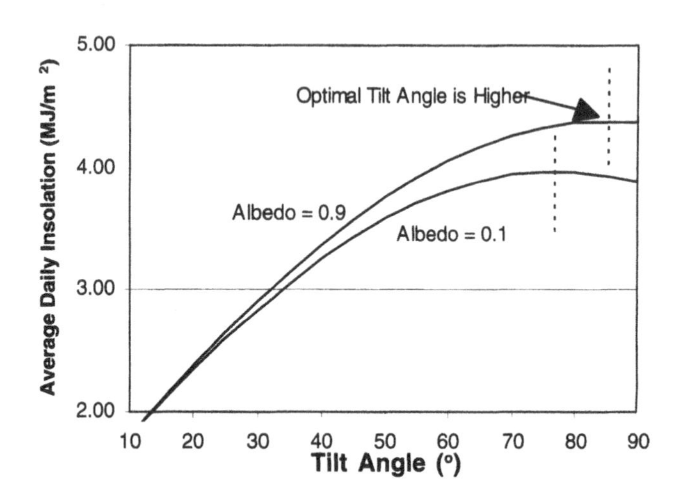

Figure 3.11. Effect of albedo on winter insolation for Stockholm, Sweden as a function of tilt angle

will be nearly as bright as the sky. In fact, when the albedo is high, higher tilt angles provide slightly more energy, as shown in Figure 3.11.

Some fixed arrays are made such that their tilt angle can be adjusted two or four times a year. Adjusting the tilt twice a year (once at each equinox) can raise *annual* output, compared with a fixed array set to the optimal angle for annual output, by only about 3 to 5%. Adjusting the tilt four times a year is not generally worthwhile – it may result in a further 0.5% improvement. These minimal improvements must be weighed against additional complexity of the array mounting structure. Adjusting the tilt angle can make sense in two cases:

- If winter output is critical and there is a use for surplus power during the summer, adjusting the tilt angle twice a year can improve annual output, compared with a fixed array optimized for *winter* production, by 10% to 25%. This is especially true at higher latitudes, where the difference between the optimal angles for summer and winter production is quite large.
- If the optimal angle for annual production is so low as to permit snow accumulation on the array, it may be desirable to raise the array tilt angle for winter. This will not increase year-round production much – the vast majority of the energy is collected outside of winter – but it will avoid heavy snow loads and provide more winter production.

ENHANCEMENT BY SOLAR TRACKING

The orientation and tilt of most arrays are fixed. This is a reliable and inexpensive approach, but it does not take full advantage of the energy available from the sun. Since the sun moves across the sky, the angle of incidence of its rays will often be quite large; ideally, the angle of incidence would be held at 0°. Trackers are structures that orient the array towards the sun in order to maintain low angles of incidence.

The additional energy produced by the array is determined by the tracking motion of the array. There are basically three tracking motions:

- *Dual-axis tracking.* The tracker adjusts both the tilt angle and the surface azimuth such that the angle of incidence of beam radiation is always kept at 0°.
- *Azimuth tracking.* The tilt angle is fixed and the tracker adjusts the surface azimuth (i.e. sweeps from east to west each day).
- *Single-axis tracking.* The array rotates around a tilted axis running through the plane of the array and intersecting the ground.

Dual-axis tracking obviously collects the most energy. At high latitudes azimuth trackers collect somewhat more energy than single-axis trackers; at low latitudes single-axis trackers work better. Azimuth trackers may be preferable in cold climates since their optimal tilt angles are higher and thus they will shed snow better.

Trackers do not significantly raise winter output of the array at mid to high latitudes. During winter at these latitudes, the sun traces a shallow arc across the southern sky (for the northern hemisphere), so there is little tracking to be done. Thus, trackers are not a solution to the problem of low wintertime insolation in cold climates. In addition, *trackers only make sense for beam radiation*: there is no point in tracking the sun when it is obscured by clouds and the whole sky is roughly equally bright. Trackers are not typically used in very cloudy climates.

Trackers can significantly improve summertime performance (e.g. by 30 to 60%), especially at high latitudes. During the summer, the sun will pass behind a fixed array; a tracker ensures that the array utilizes the long daylight period fully. At high latitudes, the vast majority of the annual solar energy is collected during the summer, so trackers can significantly improve annual energy generation as well. This is shown in Table 3.3.

In the real world, trackers do not work perfectly and collect less energy than suggested by Table 3.3. Most also consume some electric power. Chapter 5 discusses concerns about the reliability and complexity of trackers. At present, trackers are not popular in most cold climates.

ENHANCEMENT BY REFLECTORS

Another way to increase the output of the array is to place panels around it such that they reflect sunlight onto the front of the array. This can be cost-effective: the reflective panels, which are less costly per unit area than PV modules, will significantly increase the solar radiation striking the PV modules at a low marginal cost.

The wintertime performance of the panel can be significantly improved by placing a reflector *above* the PV module at an appropriate tilt angle (see Figure 3.12). One simulation suggested that during the period of November through February, a vertical array in Stockholm would receive 40% to 50% more radiation with an appropriately sized and tilted reflector.[5] The reflector, being above the array, would also reduce snow and frost on the array and the downward-facing reflective surface would remain clean. During the summer, this reflector would shade the array, drastically reducing the output, with the result that summer output would be roughly equal to winter output; this problem could be solved if the reflector was adjusted or removed for the summer.

It must be noted that the additional radiation on the array can be fully utilized only when it is illuminating the cells in the array evenly; otherwise, the situation is similar to shading. Unfortunately, reflectors will always result in somewhat uneven illumination. With thin film modules, the effects of uneven irradiation can be mitigated by turning the module such that the cells, which are typically in the form of long and narrow strips, run perpendicular to the plane of the reflector. With this arrangement, a simulation for Stockholm suggested a greater than 20% improvement in annual module output; in contrast, a conventional module would benefit by only about 10%, because of uneven illumination.

One historical problem with reflectors has been the accelerated ageing of the plastic cell encapsulant in PV modules. Better plastics have been developed and are being

Table 3.3. Theoretical increase in solar energy collected using azimuth and dual-axis trackers compared with fixed array optimally tilted for time period under study

		December		May through July		Annual	
	Latitude	Azimuth	Dual-axis	Azimuth	Dual-axis	Azimuth	Dual-axis
Miami	26°N	16%	18%	15%	22%	21%	27%
Montreal	45°N	9%	9%	28%	34%	27%	32%
Stockholm	59°N	2%	2%	55%	62%	44%	48%
Resolute	75°N	N/A	N/A	64%	66%	62%	64%

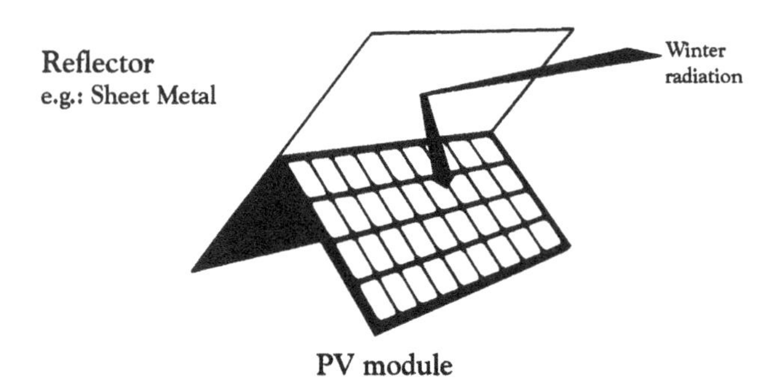

Figure 3.12. One possible design of a simple reflector that will both enhance winter radiation and reduce snow and frost problems in cold climates; in this design the reflector also acts as a module support

used now; in addition, cold climates will most probably reduce the risk of yellowing of the encapsulation material.

The extra radiation will increase the cell temperature. The module should be mounted, therefore, such that air circulation and wind around the module are maximized.

OBTAINING APPROPRIATE SOLAR RADIATION DATA

Traditionally, solar radiation data has been used mainly for meteorological and agricultural purposes. In these applications the solar radiation on a *horizontal surface* is the most important component. This kind of solar data, as well as the typical number of hours of sunshine, can be found for many locations worldwide.

Since the PV array is tilted, this is not at all sufficient for PV applications. The error will be higher at higher latitudes, where the modules have a higher tilt angle to enhance winter output. In some countries, such as Sweden, hourly measurement of solar radiation, especially for solar energy purposes, has been going on for more than a decade now and appropriate data, with separate measurements of beam and diffuse radiation, are available. Wherever such data exist, they should be used to estimate radiation on tilted surfaces.

For other cold-climate regions it may be difficult or impossible to obtain such data and the radiation on a tilted surface will have to be estimated on the basis of measurements of the global radiation on the horizontal. Often only monthly averages will be available; a number of computer programs can be used to generate hourly solar radiation data for tilted surfaces.

Weather data can be located in a variety of sources:

- *Data from maps.* Maps, such as those in Appendix A, can provide rough estimates of the solar radiation available on the horizontal anywhere in the world.
- *Data from text books and solar atlases.* Monthly averages of climatic data from all over the world can be found in many solar energy textbooks.[6] Meteorological organizations prepare atlases of solar radiation data for specific regions (e.g., the *European Solar Radiation Atlas* prepared by the Commission of the European Communities[7]). National meteorological organizations may be able to supply atlases for specific countries.
- *Data from software.* Many software packages for PV simulation and sizing include climate databases covering various geographical areas. Climate data for other places can often be bought from the software supplier. In addition, stand-alone climate data CD-ROMs are available (e.g. *Meteonorm95* from GAP Energie).
- *Data directly from Internet sources:* The Internet can be very helpful for locating suppliers of radiation data as well as radiation data itself. For example, the World Radiation Data Center maintains an on-line database located at http://wrdc-mgo.nrel.gov.
- *Satellite data.* Satellite data can be used to estimate solar radiation on a *horizontal surface* at locations where no data can be found from other sources. Also small-scale local variations in climate can be estimated this way. This is a new field that is not yet well developed for solar-energy applications; in future it may be useful for northern Asia, northern Europe and northern Canada, vast cold-climate regions containing few weather stations.[8]

REFERENCES

1. Iqbal M (1983). *An Introduction to Solar Radiation.* Toronto: Academic Press.
2. Becker CF (1979). *Solar Radiation Availability on Surfaces in the United States as Affected by Season, Orientation, Latitude, Altitude and Cloudiness.* New York: Arno Press.
3. American Society for Testing of Materials (1993). Standard E-891-87 Standard tables for terrestrial direct normal solar spectral irradiance for air mass 1.5, in the *Annual Book of ASTM Standards*, Vol. 12.02. Philadelphia, PA: American Society for Testing of Materials.
4. Wilk H (1997). Oberösterreichische Kraftwerke AG, Linz, Austria. Personal communication.
5. Rönnelid MB, Perers PK, Karlsson B (1997). Booster reflectors for photovoltaic modules at high latitudes, *Proceedings of North Sun '97, 9–11 June 1997, Espoo-Otaniemi, Finland*, pp. 555–562.
6. Duffie A, Beckman WA (1991). *Solar Engineering of Thermal Processes.* New York: John Wiley & Sons.
7. Commission of the European Communities (CEC) (1984). *European Solar Radiation Atlas.* Vol. 1, *Global Radiation on Horizontal Surfaces*, Vol. 2, *Inclined Surfaces*, W. Palz (ed.). Cologne: Verlag TÜV.
8. Ceballos JC, Barbosa De A Moura G (1997). Solar Radiation Assessment using Meteosat 4-Vis Imagery, *Solar Energy*, 60, pp.209–219.

FURTHER READING

Meinel AP, Meinel MP (1977). *Applied Solar Energy: An Introduction.* Reading, MA: Addison-Wesley.

4 The economics of photovoltaics in cold climates

Sylvain Martel and Michael M.D. Ross
(CANMET Energy Diversification Research Laboratory, Natural Resources Canada)

In cold climates, photovoltaics cannot yet compete with the relatively cheap electricity available from the grid; however, they are cost-effective for many applications located off-grid. When comparing photovoltaics with competing power sources, an economic analysis of the costs and benefits of each option must be undertaken. These costs and benefits will vary greatly depending on the solar resource available, the required reliability of the system and the remoteness of the site.

COST-EFFECTIVENESS OF PHOTOVOLTAICS

As shown in Figure 4.1, sometimes it is obvious whether or not to use photovoltaics. Consider off-grid applications that require a *small*, *safe* and highly reliable supply of *DC power* – something less than about 25 W. Here photovoltaics are almost always the most cost-effective solution: grid extension has higher capital costs and other stand-alone power systems have much higher operating and maintenance costs. PV has become the conventional solution for these applications, as shown by the many examples in Chapter 1.

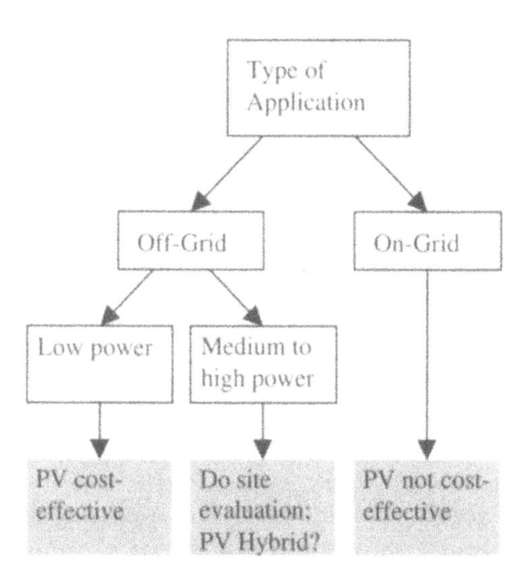

Figure 4.1. Cost-effectiveness of PV applications in cold climates

On the other hand, where electricity from the grid is readily available, photovoltaics are rarely cost-effective. There may be other reasons to use PV at these sites, such as convictions about the environment or a desire to be autonomous, but on the basis of price alone, photovoltaics are rarely justified.

In between these two there is a range of off-grid applications that require moderate amounts of power. At many sites, photovoltaics will be the least-cost solution, either in a PV-only system or a PV–hybrid system.

COSTS OF PV SYSTEMS

Capital costs for PV systems

How much does a photovoltaic system cost? That depends on the system; Table 4.1 lists 'typical' costs for three types of photovoltaic systems. The first two systems are off-grid applications, namely a residential cottage system with 150 W_p of PV modules and an industrial system for a VHF radio telecommunications site with 300 W_p of PV modules. The third system is a grid-connected, 2 kW_p rooftop-distributed system. The 'retail costs' (i.e. cost to the end-use customer) are given, both as a total cost for each system and as the cost per unit of photovoltaic capacity. These total costs include the cost of modules, inverter, controller, batteries, structure, land, design and installation. The contribution of each of these to the total cost is also shown.

Table 4.1 Typical PV system costs – in USD – in 1995 in North America[1]

	Off-grid cottage 150 W_p (USD)	Off-grid VHF telecom 300 W_p (USD)	On-grid distributed generator 2 kW_p (USD)
System cost	18.67/W_p	16.33/W_p	7.05/W_p
System cost	2800	4900	14 100
PV modules	900 (32%)	1620 (33%)	9000 (64%)
Inverter	500 (18%)	N/A	2000 (14%)
Controls	400 (14%)	500 (10%)	1000 (7%)
Batteries	400 (14%)	1080 (22%)	N/A
Structure and land	100 (4%)	300 (6%)	140 (1%)
Design and installation	500 (18%)	1420 (29%)	2000 (14%)

Table 4.1 demonstrates that modules are usually the biggest expenditure; this is especially true in grid-tied systems, which lack batteries and other components that add to the cost of off-grid systems. Design and installation is the second largest expenditure. The high cost of design and installation at the off-grid telecommunications site reflects the assumption that this system is relatively remote. As will be discussed further, remoteness generally raises costs for all types of power systems. The battery and inverter are also major costs. Battery costs are higher in the telecommunications system, partly because industrial systems such as this one use more expensive batteries than cottage systems and partly because telecommunications systems demand higher reliability, and therefore larger battery banks, than cottage systems. More information on component costs can be found in Appendix C.

Operation and maintenance costs

Photovoltaics require no fuel and little maintenance, so operation and maintenance costs are very low. In large grid-tied systems, operating costs have been around 0.005 USD per kWh of electricity generated and are expected to fall to around 0.002 USD per kWh.[2] In stand-alone systems, operation and maintenance costs are considerably higher, because of the battery. All batteries must be inspected on occasion, vented batteries must have their water replenished at least once a year and batteries have a limited lifetime and therefore require replacement (see Chapter 13). Nevertheless, these are simple tasks and capital costs are dominant with photovoltaic systems.

Life-cycle costs for PV systems

The *life-cycle cost* can be used to account for all the costs associated with a power system: it is the sum of all initial capital costs plus discounted future operating and maintenance costs, such as replacement of the batteries in off-grid systems. The life-cycle cost is an accurate basis for comparing different power sources, since it is not biased towards either capital costs or operating costs.

For the three typical systems described in Table 4.1, the life-cycle costs for a 20-year system life are shown in Table 4.2. The life-cycle costs are not much larger than the initial costs, demonstrating that for PV systems, initial costs dominate operating and maintenance costs. Regular battery replacement is assumed to be the only future expenditure for the off-grid systems and inverter replacement is assumed to be the only future expenditure for the grid-tied systems. Since PV modules have a lifetime exceeding 20 years, they do not need replacement. These life-cycle costs have been calculated using the methods described in *Solar Photovoltaic Power: The Economics*[2] assuming a discount rate of 10%, no excess inflation, no depreciation or tax benefits, that the cost of battery replacement is equal to battery cost plus one-half of the total initial installation and design cost, no costs for permits or insurance, that the salvage value equals the cost of removing equipment from the site at end of life, that the telecommunications site has sealed batteries requiring no maintenance and that the cottage owner will maintain batteries for him or herself for 'free'.

Table 4.2. Life-cycle costs for three typical PV systems

	Off-grid cottage 150 W_p	Off-grid VHF telecom 300 W_p	On-grid distributed generator 2 kW_p
Initial costs (USD)	2800	4900	14 100
Replacement	Batteries every 5 years	Batteries every 6 years	Inverter after 10 years
Life-cycle cost over 20 years (USD)	3600	6400	14 900

The costs for on-grid systems are fairly easily interpreted. With the solar resource available at mid-latitude cold climate sites, these life-cycle costs will lead to energy costs of around 0.40 USD to 0.70 USD per kWh, roughly 6 to 20 times more costly than electricity provided by conventional central electric utility grid; an excellent solar resource might halve this cost.[1]

On the other hand, the cost per unit of electricity is not that important for off-grid systems. Usually an off-grid site will have a certain power requirement and electricity generated in excess of this load, during summer for example, will be wasted. What is important is the overall cost of the system compared with competing options and this will largely depend on site-specific factors.

SITE-SPECIFIC FACTORS AFFECTING THE COSTS OF PV SYSTEMS

The size and cost of the array

For a given load, the amount of sunshine and the ambient air temperatures at the site will largely dictate the size – and therefore the cost – of the array and battery. Figure 4.2 is a *very rough* guide to the PV capacity required per watt of average load, as a function of latitude, for year-round PV-only systems, PV-only systems operating from April through September (or October through March in the southern hemisphere) and hybrid systems in which PVs power the load for six months of the year. A curve showing how the array size for a PV-only year-round system is affected by very large battery banks, capable of storing energy from the

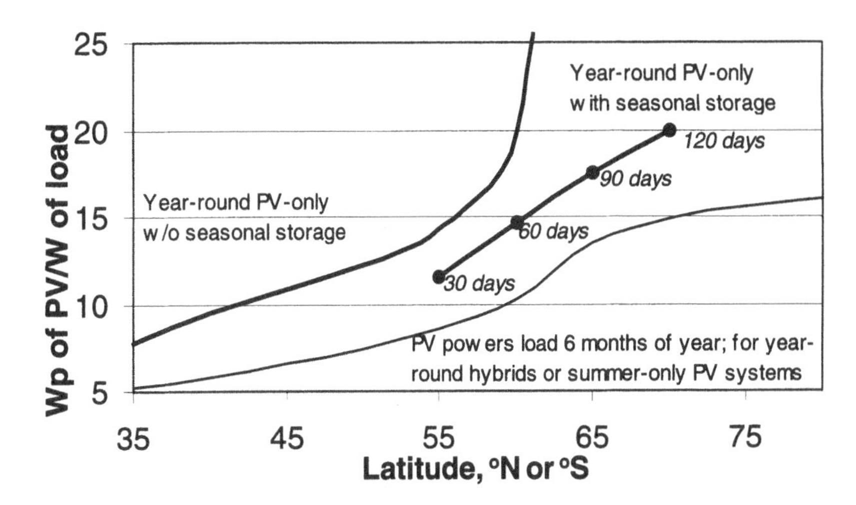

Figure 4.2. Approximate PV requirement per watt of average load over a 24 hour period

autumn into the winter ('seasonal storage'), is also shown. These estimates should be raised if the site is on a mountain or ridge prone to icing, if the site is near a coast or in a particularly cloudy area, or if shading by nearby trees or mountains is possible. Note that these curves are insufficient for design purposes, which require proper sizing methods or simulations.

Figure 4.2 illustrates several important economic considerations:

- The size of the array tends to increase at higher latitudes, confirming that costs vary from site to site.
- Sunshine is fairly 'dilute'. For every unit of load, many units of array are required. This explains why PV is most cost-effective for low-power applications: here the high cost of the array is more than offset by the benefits of a simple, low-maintenance system. Since every unit of load requires a large capital investment, PV should always be considered in conjunction with investment in *energy-efficient loads*.
- At latitudes above 55°, the size of a PV array for a year-round load escalates dramatically if *seasonal storage* is not used. In a PV system without seasonal storage, the battery is only large enough to buffer short-term variation in the availability of sunshine, that is, it transfers energy from day to day or week to week. With seasonal storage, however, the battery is large enough to transfer energy from one month to the next. This means that the array does not have to be sized to power the load fully during the darkest month or two of the year – the battery makes up the deficit. Arrays are usually more expensive than batteries, so seasonal storage can lead to lower system costs.
- Another way to avoid escalating array sizes at higher latitudes is to use a hybrid system. The array is used when sunshine is plentiful; the other generators when it is not.
- A PV-only system is much cheaper for loads that operate during the summer only. In general, system size and cost decrease whenever there is *a good match between sunshine availability and demand for power*. For this reason, it is also cheaper to power daytime loads than night-time loads.
- At a given site, the size of array required is inversely related to the size of the battery installed and vice-versa: bigger batteries permit smaller arrays and bigger arrays permit smaller batteries. Finding the optimal trade-off involves balancing the costs of the two (see Chapter 8).

The size and cost of the battery

Just as the size of the array varies from site to site, so does the size of the battery. At low latitudes and for summer-only systems, the battery buffers short-term variation only and can be fairly small. At higher latitudes, radiation levels are lower during the winter, variation affects the system more and larger batteries are required. For example, five days of storage might be sufficient for a given year-round system at 35° latitude whereas ten days might be required at 50°. At latitudes above 55°, seasonal storage demands even larger batteries. In comparison, the battery of a hybrid system buffers only short-term variation and can be quite small at any latitude – one to four days of storage is typical in PV–genset systems.

Other site-dependent factors also affect the size of the battery and therefore system cost. *Low temperatures* limit the amount of energy that can be withdrawn from a battery, inflating battery sizes. For example, at –25°C, a certain lead–acid battery may furnish only 50% of what it furnishes at 25°C. Therefore at sites where the battery temperature will drop this low, the number of days of storage must be increased by a factor of two. At mountainous sites where the array is likely to be covered in ice for extended periods of time, the battery must be further enlarged.

Reliability and costs

A final factor affecting component sizing and system cost is the desired *reliability* of the system. With a power source capable of furnishing power on demand (e.g. a genset), the reliability of the system is limited only by component failures. With a PV-only system, however, the system may fail to power the load simply because there is a period of especially cloudy weather. The probability of such a failure can be reduced, but never eliminated, through increasingly large arrays and batteries. A consequence of this is that low-reliability systems – capable of powering the load, say, 95% of the time – are relatively inexpensive, but higher levels of reliability are increasingly costly. Figure 4.3 shows that at higher latitudes, low levels of winter sunlight make reliability costly. However, even at sites like Albuquerque, where sunshine is usually strong year-round, variation in the solar resource inflates the cost of a high-reliability system.

COMPARING POWER SOURCES

Competing options and their costs

To gauge whether photovoltaics are cost-effective in a given application, they must be compared with their competition. Although grid extension, wind turbines and hydroelectric generators may be considered, for most off-grid applications PV must compete against:

- *Fossil-fuel-powered gensets.* Per unit of capacity, moderately-sized gensets are about one fifth to one tenth the price of photovoltaics; on the other hand, they require regular refuelling and maintenance and contain moving parts prone to failure. Gensets are inefficient when the demand is much smaller than the genset capacity, for example, when the genset has been sized to meet a peak load that is much higher than the average load. In these cases, batteries, a charger and a controller can be added to the genset, creating a 'cycle charger'. The battery powers the load and the genset automatically recharges the battery as required; AC loads will require an inverter in this configuration.
- *Thermoelectric generators (TEGs).* Ranging in size from 10 W to 500 W, TEGs are semiconductor devices, similar to PVs, which generate DC electricity using heat from a natural gas or propane burner. They are highly reliable when operated continuously. The purchase price is fairly high, maintenance costs are reasonably low and fuel costs depend on the availability and cost of propane or natural gas. Propane gas turns liquid at temperatures below –30°C, requiring a 'vaporizer', an additional expense.
- *Non-rechargeable batteries.* The non-rechargeable (or 'primary') battery, lacking moving parts, controls and the uncertainty of an intermittent resource, is one of the simplest and most reliable sources of DC electricity. The initial investment is very low but purchasing, transporting, installing, retrieving and recycling the batteries is expensive.

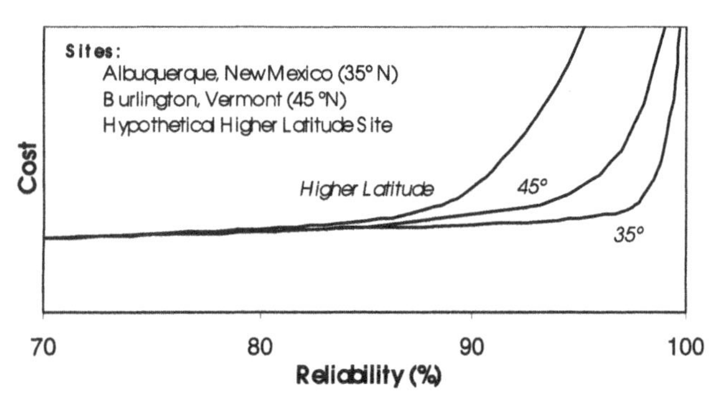

Figure 4.3. Reliability versus cost for a stand-alone PV system (adapted from Sandia National Laboratories[3])

How do these options compare? Capital costs for PV systems are relatively high, especially when compared to gensets, yet PVs can be cheaper in the long run; an economic analysis must be done to determine the costs. For some power sources, pre-project costs and initial expenses are significant; for others, recurring expenses are paramount (see Table 4.3).

Remoteness affects costs

One factor that can greatly affect the long-term cost of any off-grid power system is remoteness. Since a PV system tends to require minimal human intervention, remoteness tends to affect the costs of competing technologies, such as fossil-fuel-powered gensets, non-rechargeable batteries and grid extension, more than it affects the cost of PVs. Remoteness raises costs in five ways:

- higher fuel costs due to the cost of transportation (possibly by helicopter);
- higher costs when a system fails, since diagnosing and repairing the failure may require multiple site visits; furthermore, the load may lack power for an extended period of time;
- higher installation costs due to the expense of transport-

Table 4.3. Costs for stand-alone power sources[6]

Option	*Pre-project* Feasibility/resource assessment	Design/approvals	*Initial expenses* Installation	Equipment purchase	*Recurring expenses* Fuel	Maintenance
PVs	Low	Low	Low	Very high (Array, batteries, controls, mounting structure and inverter for AC)	Nil	Very low/Low (very low with sealed batteries; Low when battery watering and replacement needed)
Gensets/cycle charger	Nil	Low	Low	Medium (Genset, controls and batteries, plus charger if cycle charger)	Medium/High (Medium for cycle chargers)	Medium/High (Frequent inspection, fuelling, repairs, rebuilds, battery maintenance)
TEGs	Nil	Low	Very low/Low (Connection to gas)	High (TEG, control, vaporizer if cold, batteries for surge, inverter for AC)	Low/High (High unless can tap into natural gas pipeline)	Low (Occasional visit to ensure proper fuel supply)
Primary batteries	Nil	Low	Very low	Low	Very high (Battery replacement)	Medium (Annual visit for battery replacement)
Wind turbines	High (One year of data may need to be collected)	Low/Medium (Low for small, medium for large)	Low/Medium (Low for small, medium for large)	Medium/High (Turbine, tower, batteries, controls, inverter for AC; economies of scale)	Nil	Medium (Battery maintenance; occasional inspection, repairs)
Hydro	Medium/High (Costly if hydrological survey not already done)	Medium/High (Civil works design and approvals may be expensive)	Medium/High (If dam or weir is not already in place, costs may be high)	Medium (Turbine and controls, diversion load, batteries if required)	Nil	High (Frequent inspection of water path, occasional repairs)
Grid extension	Varies (Depends on length of extension and terrain)	Varies (Depends on approvals, extension length, terrain)	Medium/High (Depends on length of extension and terrain)	Low/Medium (Transformer)	Low (Reflected in low unit costs of electricity)	Nil/Very low

ing people and materials to the site or, in some cases, building access roads to the site;
- higher maintenance costs due to the cost of transportation and wages for time spent in transit;
- higher costs for grid extension.

Remoteness can make photovoltaics attractive even in seemingly unlikely applications. For example, Chapter 1 indicated that photovoltaics has a good future on remote diesel grids in Northern Canadian communities. This is in large part due to the high cost of transporting diesel fuel to these remote sites. Unlike the North American power grid, adding new peak power capacity to the local grid is relatively inexpensive – just add another cheap generator. On the other hand, fuel costs are significant and every unit of electricity generated by photovoltaics reduces fuel costs. It does not matter that the photovoltaic systems supply almost no power through the winter: they generate savings during the summer. Because of this, grid-tied PVs are almost as attractive in some of these northern Canadian communities as they are in the southern USA.[4,5] Other important factors on remote grids are the costs of transporting the PV system to the community, installation costs and the cost of balance-of-system components. Transportation and installation costs tend to be inflated by remoteness, offsetting some of the savings on fuel.

VALUING THE BENEFITS OF PHOTOVOLTAICS

Economic comparisons

There is nothing special about economic analyses for cold-climate applications. Standard methods for financial comparisons can be used to evaluate whether photovoltaics are cost-effective compared with competing options (see International Energy Agency,[7] Derrick *et al.*[2] and Jelen *et al.*;[8] Awerbuch[9] presents some more sophisticated approaches). It is worth emphasizing that all economic comparisons involving photovoltaics should:

- Consider all costs over the system's lifetime. Photovoltaics are capital intensive, while competing options are costly to operate and maintain. To determine which makes the most sense in the long term, a life cycle cost analysis should be done.
- Select appropriate rates, especially the discount rate. Since photovoltaics are capital intensive, high discount rates will make them appear less attractive and low discount rates will favour them.
- Investigate depreciation benefits. Often tax codes permit accelerated depreciation of the capital cost of part or all of a photovoltaic system. This has added significance for photovoltaics because, once again, they are capital intensive.

Intangibles

Although the economics are usually the critical indicator of the viability of a PV project, this should not be the final word because they do not reflect all the benefits associated with the use of photovoltaics. Intangible considerations are often as important as costs and include:

- *Risk mitigation.* The true life-cycle cost of a power system is determined by a large number of variables that are unknown at the time the power system is purchased. These unknowns affect PVs less than competing power sources, lowering the risk of the system costing more than projected:
  - No fuel is used, so escalating fuel costs are not a concern.
  - PVs are simple and highly reliable, minimizing the risk of expensive repairs and down time.
  - Unlike grid extension, a PV system can be transported to another site if the load is eliminated.
  - Unlike gensets or TEGs, PV systems can expand to accommodate future changes in demand.
- *Management and administrative costs.* Operating and maintaining a PV system is simple, so management and administrative costs are low. Furthermore, the solar resource is much less site-dependent than other renewables, so a standard design or family of designs can be used for all sites within a region.
- *'Green' publicity.* Properly publicized, PV systems enhance a company's image much more than is warranted by their environmental benefits. This is not cynical; today's markets for PV encourage research and development that will result in falling costs for PVs, larger installations and correspondingly larger environmental benefits in the future.
- *Local pollution.* Photovoltaics generate neither noise nor noxious emissions. This can improve living and working conditions for employees and avoid the public's illwill.

Externalities

Unlike a company, which is concerned only with maximizing the benefits accrued to itself, government, public utilities and some not-for-profit organizations exist to maximize the benefits for all of society. As such, they should base decisions not just on an analysis of the immediate benefits and costs, but also on the 'externalities', or costs and benefits accrued to society as a whole.

Photovoltaics do less damage to the natural environment than non-renewable technologies. They emit no particulates or greenhouse gasses during operation and the risk of a fuel spill is lowered or eliminated. Frequent site visits are not necessary, so erosion-causing access roads need not be built. An environmental criticism sometimes levelled at PVs is that they require more energy in manufacture than they generate during their lifetime. This is simply false. In cold-climates, the energy payback time of a photovoltaic module is one to five years, leaving 15 to 25 years for net energy generation.[10]

It is difficult to assess the value of these externalities in stand-alone installations, but they can be significant. For example, a 300 gallon diesel fuel spill in a US park, which can occur during transport of fuel to a genset, can cost more than 100,000 USD to clean up.[11] For large-scale generation on the grid, the pollution prevention benefits of PVs have been estimated at 0.003 USD to 0.010 USD per kWh versus natural gas, 0.022 USD to 0.069 USD per kWh versus oil, 0.022 USD to 0.059 USD per kWh versus coal and 0.026 USD to 0.030 USD per kWh versus nuclear.[12]

REFERENCES

1. Leng G, Dignard-Bailey L, Bragagnolo J, Tamizhmani G, Usher E. (1996). *Overview of the Worldwide Photovoltaic Industry.* Varennes, Québec: CANMET Energy Diversification Research Laboratory

2. Derrick A, Paish OF, Fraenkel PL, Louineau JP (1995). *Solar Photovoltaic Power: The Economics*. Eversley, Hants, UK: IT Power Ltd.
3. Sandia National Laboratories (1995). *Stand-alone Photovoltaic Systems: A Handbook of Recommended Design Practices*. Albuquerque, New Mexico: Sandia National Laboratories.
4. Martel S, Usher E (1995). The value of using photovoltaics to displace fossil fuel consumption of NWT diesel-electric grids. *Proceedings of the 21st Annual Conference of the Solar Energy Society of Canada, Inc.* Ottawa: SESCI.
5. Osborn D (1997). Commercialization of utility PV distributed power systems. *Proceedings of the 1997 Annual Conference American Solar Energy Society, April 1997, Washington, DC.* pp.153–159. American Solar Energy Society, Boulder, Colorado, USA.
6. Leng G, Ah-You K, Painchaud G, Meloche N, Bennett K, Carpenter S, Thevenard D, Gordon J, Hollick J, McCallum B, Price C, Lodge M, Leenders E, Graham S, Marshall R, Sellers P (1998). *RETScreen – Renewable Energy Technologies Project Assessment Tool Version 98*. Varennes, Québec: CANMET Energy Diversification Research Laboratory. http://cedrl.mets.nrcan.gc.ca/retscreen/
7. International Energy Agency (1991). *Guidelines for the economic analysis of renewable energy technology applications: based on the findings of the International Energy Agency Workshop on the Economics of Renewable Energy Technologies*, Château Montebello, Quebec, Canada. Paris, France: International Energy Agency.
8. Jelen F, Black J (1978). *Cost and Optimization Engineering*, 2nd edn. New York: McGraw-Hill.
9. Awerbuch S (1996). New economic perspectives for valuing solar technologies. *Advances in Solar Energy*, Vol. 10. New York: American Solar Energy Society.
10. Nijs J, Mertens R, Van Overstraeten R, Szlufcik J, Hukin D, Frisson L (1997). Energy payback time of crystalline silicon solar modules. *Advances in Solar Energy*, Vol. 11. New York: American Solar Energy Society.
11. Post H (1998). Sandia National Laboratories, Albuquerque, New Mexico. Personal communication.
12. Chupka M, Howarth D (1992). *Renewable Electric Generation: An Assessment of Air Pollution Prevention Potential*. Washington, DC: United States Environmental Protection Agency.

II The effects of cold climates on photovoltaic systems

5 The array, structure and electronics

Heinrich Wilk
(Oberösterreichische Kraftwerke AG)

Photovoltaic arrays, mounting structures and electronic components are affected by the low temperatures, snow, ice and darkness of cold climates. Some of these effects are positive; others are negative. Problems can be avoided if the impact of the climate is given due consideration during design.

PHOTOVOLTAIC MODULES IN COLD CLIMATES

Photovoltaic modules perform slightly differently in cold climates. Cold cell temperatures, low light levels, altered light spectrum, high incidence angles of the sun's rays and snow and ice accumulation all affect the module's operational characteristics. Cold cell temperatures are probably the most important consideration; fortunately, cell efficiency improves at cold temperatures.

Cold cells work better

When sunlight strikes a photovoltaic module, some of the energy is transformed into electricity, but most of the energy just heats up the module. As a result, it becomes hot: under bright sunlight, the cells of a typical module will be 25 to 30°C warmer than the surrounding air.

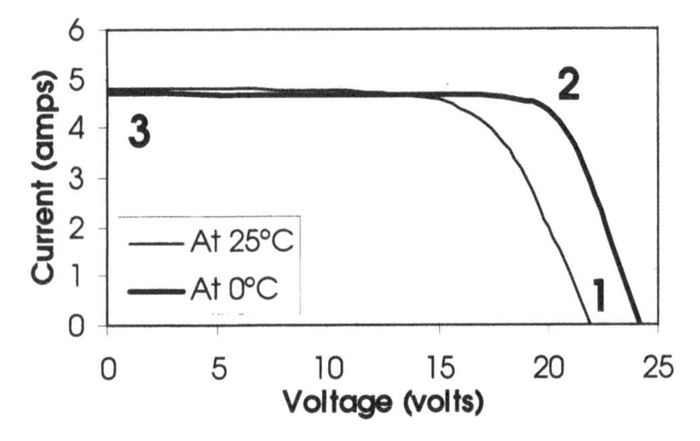

Figure 5.1. Effects of low cell temperatures on the *I–V* curve of a crystalline silicon PV module: 1. The voltage rises; 2. The fill factor improves; 3. The current drops slightly

Most photovoltaic cells work better when cold. At low temperatures, the electrons in the cell are less agitated and, as a consequence, the photovoltaic effect operates more efficiently. This improves the power output of many popular photovoltaic technologies.

Low temperatures change the electrical characteristics of the module in three ways, as illustrated in the *I–V* curve of Figure 5.1.[1] First, at a given light level and current, the voltage will be higher, except at currents approaching the short circuit current. Most noticeably, the open circuit voltage V_{oc} rises. Since power is the product of current and voltage, raising the voltage leads to more power for a given light level. Second, the shape of the *I–V* curve changes slightly, so that it appears squarer. This improvement in the 'fill factor', or the squareness of the curve, translates into higher output when the module is operating near the maximum power point. Third, currents are shifted very slightly downward, as evidenced by the change in the short-circuit current, I_{sc}. Usually the drop in current is negligible.

The peak power and open-circuit voltage of most types of modules tend to improve by around 0.3% to 0.5% for every degree Celsius drop in cell temperature. Among common types of modules, only amorphous silicon shows little or no improvement at low temperatures. In fact, relatively high cell temperatures (above 60°C) actually result in better performance for this technology, owing to the peculiarities of its physics.[2]

Improvements in the *I–V* curve translate into better real-world performance. Grid-tied systems and battery charging systems with maximum power point tracking will produce proportionally more electricity. Maximum power point trackers (MPPT) are electronic controllers that keep the array operating at its maximum power point, regardless of the operating voltage of the circuit that is being powered by the array, e.g. regardless of the battery voltage. They have nothing to do with physically tracking the sun.] For example, a crystalline silicon array will generate roughly 10% more electricity in a climate averaging 0°C than in one averaging 25°C, given that the climates are equally sunny. On the other hand, battery-charging systems without maximum power point tracking will not see much improvement, since the battery voltage fixes the voltage for the whole system, including the photovoltaic array. If the module voltage is fixed by the battery at, say, 13 V, then no benefit is derived from higher open-circuit voltages and maximum power points. There is one way to take advantage of higher voltages, however, and that is to reduce costs by using modules with fewer cells. This lowers module voltages and consequently module power. In warm climates, a module with fewer cells may not be able to operate at a voltage high enough to charge the battery. In cold climates this problem is unlikely and modules with 30 or 33 cells in series can be used in lieu of those with the normal 36. Before selecting such modules, however, the designer must ensure that the climate is sufficiently cold to achieve full charging at all times of the year.

Poor performance at low light levels

In many cold-climate regions, winter days are darker than the days of other seasons. Weather tends to be cloudier and, at moderate to high latitudes, the sun's rays often strike the photovoltaic array at a high angle, decreasing the amount of radiation that reaches the modules. As a result, cold-climate PV systems may spend many hours illuminated by weak sunlight. Most modules are less efficient under weak lighting: when irradiated by, say, 200 W/m², they generate less than 20% of the power that they would produce under bright sunlight of 1000 W/m². This diminishes winter output slightly.

Not all modules are created equal, even when they perform identically under standard test conditions (that is, 1000 W/m² and a cell temperature of 25°C).[3–6] The efficiency of three hypothetical modules is graphed as a function of irradiance in Figure 5.2. Module A is typical: when irradiance drops below 200 W/m², efficiency falls rather precipitously. Module B is representative of an especially poor module; for some modules, efficiency falls 30% by 200 W/m².

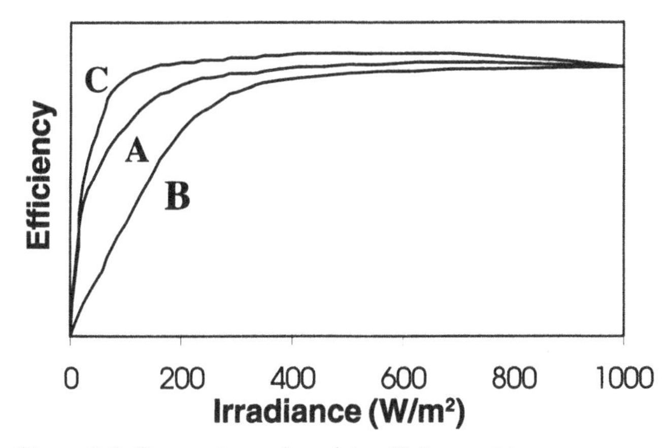

Figure 5.2. Comparison of module efficiency at low irradiance levels. A = Typical, B = Poor low light performance, C = Good low light performance

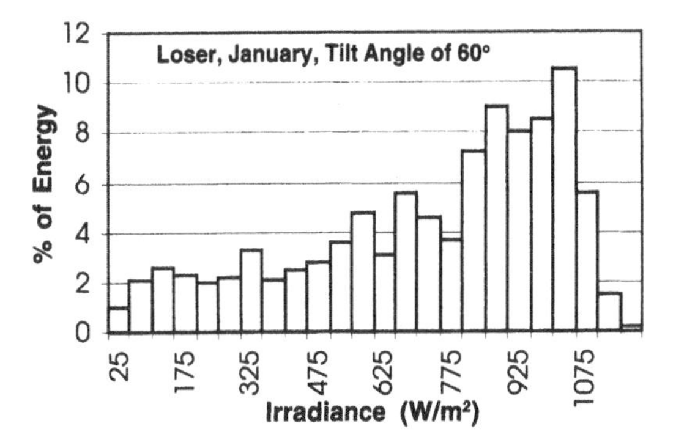

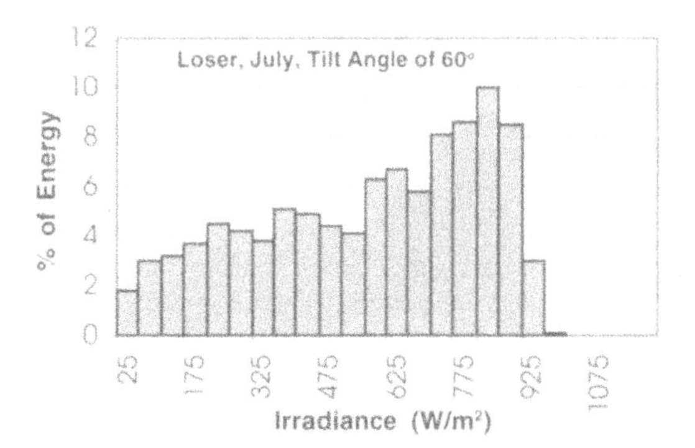

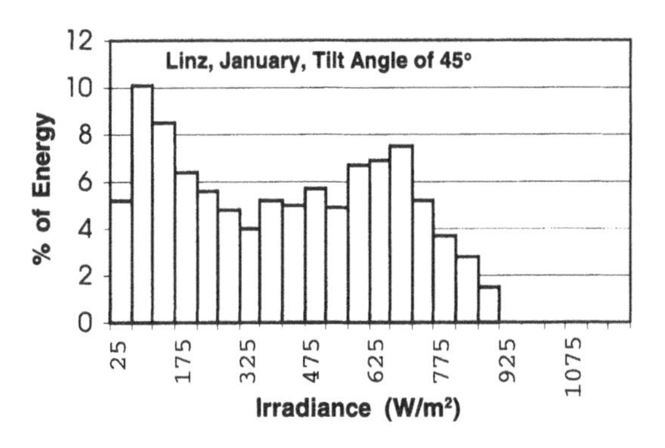

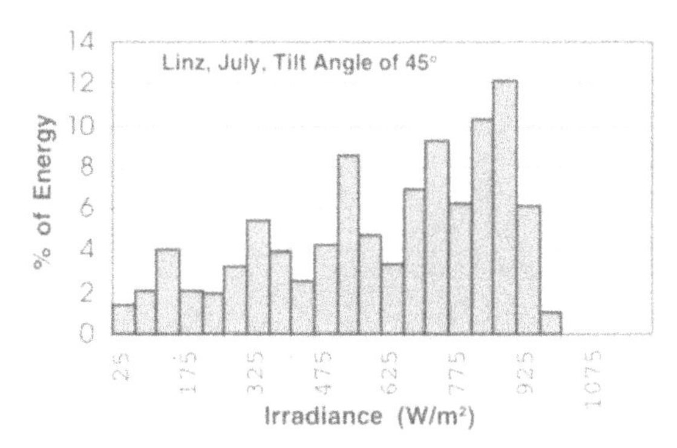

Figure 5.3. Distribution of solar radiation at Mount Loser (1550 m) and Linz (350 m), Austria

Modules like Module C, with relatively good efficiencies at low light levels, are rare.

The effect of poor low-light efficiency on a real photovoltaic system depends on the fraction of the total sunlight that is received at low light levels. Some cold-climate sites are characterized by bright sunshine most of the time. For example, the PV system installed at 1550 m altitude on Mount Loser in the Austrian Alps receives most of its radiation at relatively high irradiance levels, even in winter. In contrast, for a system installed at 350 m altitude in Linz, Austria, low light levels are the norm during winter (Figure 5.3).

Surprisingly, for both Mount Loser and Linz, poor efficiency at low light levels will lower annual energy production by only 3% to 4% for typical panels. Why is the difference between Mount Loser, which has high irradiance levels year-round, and Linz, which has a dark, foggy winter, not more pronounced? The key is *annual energy production*. In Linz, so much more energy is produced in summer than in winter that poor low-light performance during winter is masked. Grid-tied systems at either Mount Loser or Linz would be minimally affected by poor low-light performance.

On the other hand, low-light performance could be critical for a year-round stand-alone PV system installed at Linz, which would have to be designed to power the load through the worst-case winter months. During January, 40% of the total solar energy received is at irradiance levels below 300 W/m^2. A poor module (Module B) would produce roughly 20% less energy than an exemplary module (Module C). To keep the stand-alone system from running into an energy deficit for January, an array 25% larger would be required with Module B than Module C; otherwise, reliability would be compromised. Low-light performance can also be significant for systems at higher latitudes, where higher sun angles may cause a fair portion of system energy to be collected at low irradiance levels, year-round.

At present, it is difficult for the consumer to differentiate between modules with good and poor low-light performance. Manufacturers specify performance under standard test conditions only. There is movement, however, towards specifications that would include module output at a range of irradiance levels, facilitating the selection of modules with high efficiency at low light levels for those sites that are often cloudy.

Spectral effects

As discussed in Chapter 3, sunlight contains a full range of colours. During winter at moderate and high latitudes, the sun is low in the sky and its light must pass through more air. This tends to scatter the blue light and the sun's rays become relatively more red as a consequence. This has little effect on crystalline silicon modules, since the modules are sensitive to a wide range of wavelengths of light. Amorphous silicon and cadmium telluride modules are sensitive to a much narrower band of colours, however, and the winter shift to redder sunlight results in slightly poorer performance. At moderate latitudes, a wintertime penalty of about 10% can be expected with amorphous silicon; higher latitude installations may suffer more.

Snow and ice accumulation on the array

In cold climates, snow falls and water freezes. Sometimes snow and ice can accumulate on the array, blocking sunlight, reducing the electrical output, and in some rare cases, damaging the modules. Mountainous areas susceptible to atmospheric icing are particularly troublesome. Fortunately, most installations have few such problems.[7]

The most obvious manifestation of a cold climate is snow, but other forms of frozen or freezing precipitation occur. Freezing rain and freezing fog tend to envelop surfaces and structures in a transparent ice called 'glaze'. 'Frost' or 'hoarfrost' occurs when water vapour in the air forms ice crystals on a surface; it is the translucent accretion that must be scraped off car windshields in most cold climates. 'Rime' is an opaque white ice that occurs along ridges and near the summits of some mountains. It results when humid air, such as is found in the vicinity of an open lake, river or ocean, encounters a mountain, rises and consequently cools. As it cools, tiny water droplets form. If the air is relatively clean, the droplets can 'supercool', attaining temperatures well below 0°C while still remaining liquid. When these droplets touch a surface, such as a photovoltaic array, they freeze. Like frost, rime adheres well to glass; it is usually denser and more opaque than snow.

Snow, glaze and hoarfrost are rarely serious problems. They are all relatively transparent, so sunshine can pass through them, warming the photovoltaic array beneath them. The array soon melts the snow or ice that is in contact with the array, creating a layer of lubricating water. If the array is at a reasonable tilt angle, the snow or ice will slide off by gravity; otherwise it will simply melt away. Figure 5.4 shows snow sliding off an array at Linz. On an annual basis, snow accumulation results in a loss of only about 3.5% of total potential energy production at the Mt. Loser site; snow causes losses of only 1% to 1.4% at lower altitude sites in Austria.[8] In most years, however, there is at least one month during which losses due to snow accumulation at the Mt. Loser site are on the order of 15%.

In contrast, rime can plague a mountainous photovoltaic system for much of the year, requiring enormous battery banks for those periods when the array is encased in ice. There have been reports of photovoltaic modules covered in

Figure 5.4. Snow melting off a PV array at Linz, Austria

rime accumulations exceeding one metre in thickness. Since little sunlight can penetrate a layer of this thickness, it is not quickly melted and the photovoltaic system will fail if the battery is not large enough to supply the entire electrical requirement for a period of up to a month or more. It must be emphasized that rime is a problem only at certain sites and that its severity varies from year to year. Open water nearby and brisk winds are prerequisites; it tends to occur where there is an abrupt rise in elevation exceeding 500 m. Rime may also occur in Arctic and Antarctic regions prone to fog.

Under extremely rare circumstances, snow and ice accumulation can damage photovoltaic modules. There is one report of very heavy accumulations of ice and snow causing stress sufficient to crack module glass. There are several reports of heavy ice and snow accumulations deforming the aluminium module frame; it is speculated that the damage occurred as a result of the weight of the accumulation on the bottom part of the frame when the snow and ice slid down the module. Such problems are generally unheard of except at the most difficult alpine environments. Modules may also be damaged by ice falling from nearby structures and guy wires.

Expensive countermeasures, such as drastically increasing the size of the battery bank, need be considered only at sites prone to rime icing. Snow and ice accumulation can usually be minimized in a number of ways:[7]

- *Higher tilt angles* increase the probability of the snow sliding off by gravity. When winter performance is important, tilt angles should be at least 45°; in areas subject to very heavy snowfall, angles of 60° or more are recommended.
- *Mounting rectangular modules with their shorter dimension parallel to the ground* (i.e. as the modules are mounted in Figure 5.4) encourages snow to slide off the array. Horizontal sections of the module frame provide a foothold for snow accumulation; the longer the run of glass above the bottom section of the frame, the more likely the snow at the bottom of the module will be unable to support the weight of the snow above it. For similar reasons, *small modules tend to retain snow longer than large modules*.
- *Selecting sites that are out of the wind* if rime is a concern. Winds carry the water droplets that cause rime; situating a PV system along a windy ridge or on an exposed mountain top will exacerbate the problem.
- *Using by-pass diodes and horizontal strings of series-connected modules* minimizes the electrical consequences of partial shading by snow and ice. When cells are connected in series, partial shading of the module will severely curtail electricity generation: all the current that is being generated by the unshaded cells must be forced through the shaded cells. This causes heating of the

shaded cell, which accelerates aging. To prevent this, modules should have built-in 'by-pass diodes', which permit current to flow around shaded sections of the module. Analogously, modules connected in series will perform poorly even when only one module is shaded. Since snow and ice tend to melt off photovoltaic arrays from the top down, series-connected modules should be grouped in horizontal rows, as opposed to vertical columns, whenever possible.

- *Careful selection or modification of the module frame* can improve snow shedding and eliminate the possibility of frame damage. Laminates completely lacking a frame obviously shed snow best; sometimes removing just the bottom part of the frame can help. Even the heaviest snow and ice accumulations will not damage certain especially strong frames, such as that on the Siemens SM50 module produced in Munich, Germany, up to 1994. Standard aluminium frames can be reinforced with angle iron. For 'comshells', silo-like towers which house telecommunications equipment in mountainous areas, frameless modules that mount nearly flush with the surface of the comshell have been developed. It has been reported that this reduces accumulation.
- *Manual removal* of snow and ice is an option when people regularly visit the site. However, it is time-consuming, introduces the danger of a person accidentally damaging the module and yields very little additional energy in most cases. It may be justified when module damage is a concern or when rime is problematic.
- *Passive melting technologies*, which use the sun's heat to accelerate melting of snow and ice, are available and work well for snow. It is not clear how well these work for rime ice.
- *Active melting*, by reversing the flow of current through the array, causing it to heat up, has been investigated experimentally at Oberösterreichische Kraftwerke AG. With no wind and temperatures near freezing, five centimetres of snow were melted off two Kyocera 48 W panels in three hours with a DC current of 2 A at 48 V. Thus, melting off the snow required the equivalent of the output of the photovoltaic system during three hours of full sunshine.
- *Ice-phobic coatings*, which impede adhesion of ice to glass, appear to have been unsuccessful in the field.

STRUCTURES

Constraints on the design

Cold climates impose two constraints on the design of the array support structure:

- The structure must elevate the bottom of the panels above the surface of the snow even in worst-case conditions. Just because a PV system is installed in summer does not mean that the designer can forget about winter! Some PV systems have been buried in snow during winter. Note that more snow will accumulate under the array than in surrounding areas both because snow that slides off the array will form a pile at the base of the array and because the array will act as a windbreak, encouraging snow deposition on the leeward side of the array.
- The structure must be able to survive the combined maximum load of snow, ice and wind. If in doubt, consult a structural engineer.

Tracking systems in cold climates

Arrays that track the position of the sun in the sky can increase the amount of energy produced by the array. Both 'active' and 'passive' trackers exist; an active tracker contains a device sensing the sun's position and a motor to adjust the array position, while a passive tracker typically employs gravity and the movement of a fluid within a cylinder to turn the array. In either case, the system designer must determine whether the tracker is worthwhile: that is, does the increased energy production and consequent reduction in array size justify the additional cost and complexity of the tracker? The first part of this question – how much more electricity would be generated with the tracker – was discussed in Chapter 3. This leaves the topic of complexity and reliability.

A well designed tracker is a fairly reliable device, but small problems may arise. Passive trackers can be blown off course; active trackers can blow a fuse, have their sun sensor covered by snow or bird droppings or even have a motor burn out. Because they introduce additional complexity to the system, trackers are rarely employed in critical standalone applications, such as telecommunications repeater stations, especially in cold climates. For accessible, noncritical loads, such as remote residences and summertime water pumping, added complexity is not a major issue.

Cold climates create additional difficulties for trackers. Active trackers rely on motors and gears that may have a tendency to jam at low temperatures. This can blow fuses and prevent the array from turning. During summer at high latitudes, the sun may rise and set so far 'behind' the array that an active tracker can become confused; a mechanism to return the array to an eastern orientation at the end of each day, ready for sunrise, may be necessary. Some passive trackers require that the array be mounted at a low tilt angle, encouraging snow accumulation and diminishing winter energy generation. Further winter losses occur when a passive tracker 'falls asleep' in a westerly position and, owing to the combination of low temperatures and weak sun, does not start tracking the sun properly until late in the morning.

In cold climates, many system operators turn off their trackers in the winter and fix the array so that it faces due south (or due north in the southern hemisphere). This diminishes reliability issues, eliminates the electrical drain of the active tracker and allows the array tilt angle to be raised beyond what is normally possible with a passive tracker. The performance penalty is minimal, especially at moderate to high latitudes.

Abrasion by blowing snow and ice

At especially cold and windy sites, blowing snow and ice can be highly abrasive, stripping paint and eroding soft surfaces. Anodized structures may outlast painted structures in these environments. Some installers actually cover the rear face of the module, although the need for this has not been demonstrated.

ELECTRONICS IN COLD CLIMATES

Photovoltaic systems usually include some electronic equipment, typically a charge controller and an inverter. Cold climates can affect these electronic components in a number of ways.

High peak current and power

At certain times of the year, most cold climates are simultaneously cold and very bright. This combination of conditions is ideal for the photovoltaic array and it will produce far more power than would be expected in a warmer climate. For example, at the Mount Loser alpine PV system, highly reflective snow cover on the ground leads to irradiance levels of 1200 W/m^2 (see Figure 5.3), producing array currents roughly 20% higher than specified for the array under standard test conditions. Furthermore, these high currents occur in winter when cell temperatures are low so array voltages – and power – will be higher.

The charge controller must be able to handle these peaks in current and power. Battery charge controllers should be specified for currents about 30% higher than would be produced by the array under standard test conditions; those controllers that track the maximum power point should use an even larger safety factor.

These peaks are not pertinent to the sizing of the inverter. In stand-alone applications, inverters are matched to the load, not the array; though array output influences the choice of inverter in grid-tied applications, oversizing is rarely beneficial. Peaks in the array power will not damage most inverters: if they cannot handle the peak, they just move away from the maximum power point such that the array produces less power. Furthermore, inverters operate less efficiently at low power levels: oversizing decreases efficiency, especially for the low irradiance conditions that are common in many cold climates.

Low temperatures and condensation affect electronic components

Many of the electronic devices within charge controllers and inverters are sensitive to temperature. A few devices fail at low temperatures; for example, liquid-crystal displays are often unreadable and some buttons and other input devices may not function. While most devices work over a wide range of temperatures, the characteristics of diodes, transistors and capacitors can shift slightly when cold. When all these subtly altered devices are integrated into a circuit, the result may be a charge controller or inverter that works, but with specifications quite different from those at 25°C.

It is not just the cold that can affect electronics: cold climates may exacerbate thermal cycling, encouraging condensation. Condensation can infiltrate certain devices, such as capacitors, causing drift in the device's specifications. It can also short circuit components operating at voltages of about 70 V or higher.

Points to consider during component selection

When selecting a charge controller or inverter for a cold climate, several points must be considered:

- *Specifications.* According to the specification, does the component function over the necessary range of temperatures? And, if it does function when cold, will its performance remain constant? Many charge controllers operate properly, hot or cold, but specifications indicate that most inverters either malfunction or operate poorly at temperatures below 0°C.
- *Simplicity.* As a general rule, simple circuits are more reliable at cold temperatures than complex circuits. Circuits that operate at relatively high frequencies (say, 40 kHz or higher) will have to be very carefully designed if they are to operate when cold. They will also tend to produce electromagnetic noise that may interfere with nearby telecommunications or electronic equipment.
- *Devices.* As mentioned above, liquid-crystal readouts and some buttons work poorly when cold. It may be necessary to avoid these. The characteristics of other devices shift with temperature. If this is a concern, circuits built with 'military specification' devices may outperform those employing 'commercial' devices.
- *Potting.* Some electronics are completely encased in epoxy. This 'potting' protects the circuit and avoids condensation. Certain types of potting may stress the electronic

components during thermal cycling, since the epoxy will have a different coefficient of thermal expansion from the components. Many types of epoxy exist; ensure that the potted circuit has been cycled over a range of cold and warm temperatures.

- *Fans.* Fans, which may be part of a circuit's cooling system, may fail at low temperatures. Components that are cooled by free convection and radiation are preferable.
- *Plastic cases.* Some plastics are fragile at low temperatures.

Insulating electronic components

Ideally, electronic components should be kept in a dry environment at a fairly constant temperature between 10 and 25°C. If a building near the site offers such an environment, consider putting the inverter and possibly even the charge controller in the building. When such an environment does not exist, the electronic components are sometimes placed in an insulated enclosure. Exercise caution: these components can generate considerable heat and may overheat during summer. In addition, these components should not be placed with the batteries, since batteries may generate explosive hydrogen gas that can be ignited by sparks in the electronic equipment. In addition, corrosive vapours from the battery could attack electronic equipment.

REFERENCES

1. Rauschenbach HS (1980). *Solar Cell Array Design Handbook.* New York: Van Nostrand Reinhold.
2. Emery K (1995). Temperature and irradiance behavior of photovoltaic devices, in *Proceedings of the Photovoltaic Performance and Reliability Workshop, September 7–8 1995, Golden, Colorado*, L. Mrig (Ed.). Golden, CO: National Renewable Energy Laboratory.
3. Bücher K (1995). True module rating: Analysis of module power loss mechanisms, *Proceedings of the 13th European Photovoltaic Solar Energy Conference, October 23–27 1995, Nice, France*, pp. 2097–2103. H.S. Stephens & Associates, Felmersham, Bedford, UK.
4. Mason NB, Bruton TM, Heasman KC (1997). Factors which maximise the kWh/kWp performance of PV installations in northern Europe, *Proceedings of the 14th European Photovoltaic Solar Energy Conference, 30 June–4 July, 1997, Barcelona, Spain*, pp. 2021–2024. Published on behalf of WIP, Munich by HS Stephens & Associates, Felmersham, Bedford, UK.
5. Nagel DI (1993). DASA Wedel, personal communication.
6. Wilk H (1997). Electricity yield of PV-systems in different climates and dependence of module efficiency as a function of irradiance and other factors, *Proceedings of the 14th European Photovoltaic Solar Energy Conference, 30 June–4 July, 1997, Barcelona, Spain*, pp. 297–300.
7. Ross M, Usher E (1995). Photovoltaic array icing and snow accumulation: a study of a passive melting technology. *Proceedings of the 21st Annual Conference of the Solar Energy Society of Canada, Inc., 31 October–2 November 1995, Toronto, Canada*, pp. 21–26. Solar Energy Society of Canada, Inc., Ottawa, Ontario, Canada.
8. Wilk H, Szeless A, Beck A, Meier H, Heikkilä M, Nyman C (1994). Eureka Project EU 333 Alpsolar: Field testing and optimization of photovoltaic solar power plant equipment, Progress Report 1994, *Proceedings of the 12th European Photovoltaic Solar Energy Conference, 11–15 April 1994, Amsterdam, The Netherlands*, pp. 854–858. HS Stephens & Associates, Felmersham, Bedford, UK.

6 Battery issues

David J. Spiers
(Neste Advanced Power Systems UK)

The output of a PV array depends on the availability of sunlight. Such a variable output is suitable only for a few applications, such as water pumping or in grid-connected systems. In most cases loads require that electricity is available on demand. This is especially significant in cold climates, where sunlight can be minimal for long periods of time. Consequently, PV systems used in remote cold locations use storage devices, the most common being batteries.

Batteries provide three important functions in a photovoltaic system:

- *autonomy*, by meeting the load requirements at all times even at night or during overcast periods;
- *surge-current capability*, by supplying, when necessary, currents higher than the PV array can deliver, especially to start motors and other equipment;
- *voltage control*, thus preventing large voltage fluctuations that may damage the load.

In remote cold locations, batteries are extremely important because they are often the only reliable means of powering the loads during long overcast or sunless periods. For this reason, the battery bank is usually a large component of the power system and special attention must be given to its design and operation.

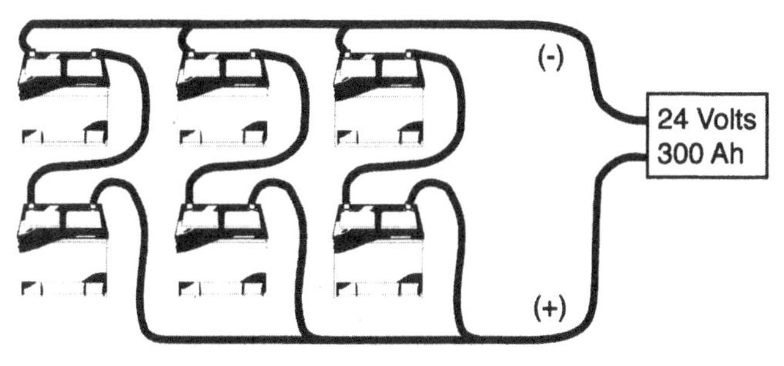

Figure 6.1. Series–parallel connection of batteries

GENERAL

Battery basics

A battery is composed of a number of electrochemical storage cells, which store electricity in the form of chemical energy. When a cell is connected to an electrical load (or circuit), a chemical reaction occurs between the 'poles' (or 'electrodes') of the cell, which results in a flow of electrons through the cell to the load. At the positive electrode, the active material undergoes chemical reduction, absorbing negatively charged electrons from the external circuit. At the negative electrode, the active material undergoes chemical oxidation, releasing electrons into the external circuit. Between the two electrodes, electric charge is transported through the cell by ions (e.g., H^+, O^-) in an aqueous solution (the electrolyte) to complete the electrical circuit. The process of the cell releasing its stored chemical energy into useful electrical energy is called discharging.

In *primary cells*, the chemical processes that occur during discharging cannot be reversed by applying an external electrical current. In fact, the result of trying to recharge such cells is often a dangerous build up of gas inside and quite possibly an explosion. Examples of primary cells and batteries are the familiar 'dry cells' and alkaline cells used in many consumer products, as well as the more specialized button cells and batteries used in hearing aids, watches, calculators, cameras, etc.

In *secondary cells* and batteries, also described as rechargeable batteries, storage batteries or accumulators, the chemical processes of discharging can be reversed by applying an external electrical current.

This process is called 'recharging', or simply 'charging'. The charging process has to be carried out with some care, especially to prevent damage from overheating or overcharging. Examples of rechargeable battery systems are lead–acid, nickel–cadmium, nickel hydride, nickel–iron and rechargeable lithium types. The two most commonly used rechargeable battery types in photovoltaic systems are lead–acid and nickel–cadmium; the applicability of other battery types to PV applications has not been demonstrated and they will not be discussed in this book.

When no external circuit is connected, the cell is said to be at open circuit. A voltage exists between the two electrodes which reflects the difference in chemical potential energy of the two active materials. The nominal open-circuit cell voltages are 2 V for lead–acid, 1.2 V for nickel–cadmium and around 1.5V for zinc–carbon (the common non-rechargeable 'dry' cells) and alkaline manganese cells (the longer-lasting non-rechargeable 'alkaline' cells). These cell voltages are determined by the chemical nature of the active materials.

In order to achieve higher voltages, cells are connected together in series to form a battery. This connection may be made by assembling cells together in one battery package or it may be made by external interconnections between individual cells (Figure 6.1). In either case, the assembly of cells is called a battery. Batteries may also be connected in parallel to increase the capacity (ampere-hours) of the battery. Interconnected groups of batteries are usually called battery banks.

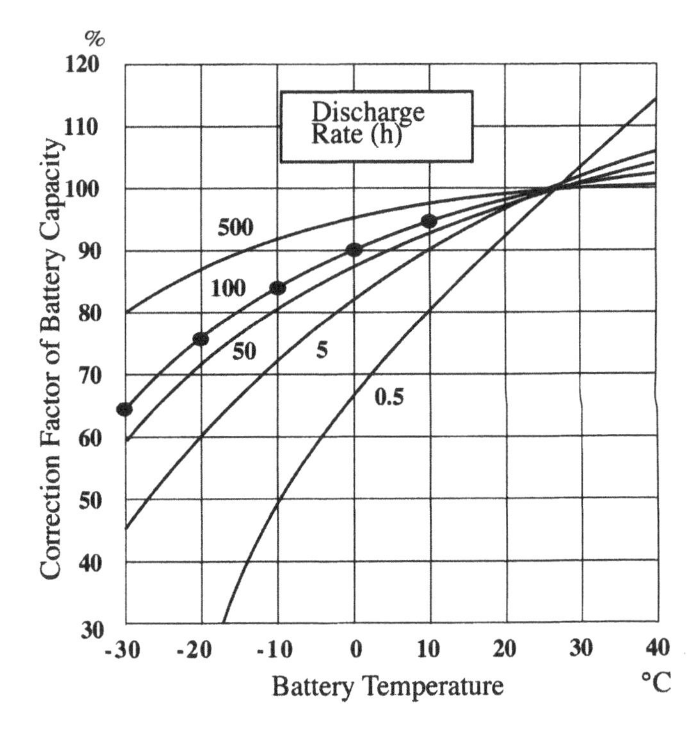

Figure 6.2. Typical lead–acid battery capacity at different rates of discharge and temperatures (note that this graph does not show that at low states of charge the battery may freeze before reaching the low temperature or state of charge shown here. See Figure 6.7 and associated text)

Battery capacity

Battery capacity is usually specified in ampere-hours (Ah) and should indicate how long the battery could operate until it reaches a specific cut-off voltage under a set load and without any recharge. Theoretically, a 100 Ah battery should deliver either 100 A for one hour, 50 A for 2 hours or 5 A for 20 hours before reaching its cut-off voltage. This is not really the case since if the battery is charged or discharged at a rate different from the one specified, the available ampere-hour capacity will vary. Generally if the battery is discharged at a low rate, its capacity will be higher (Figure 6.2).

Another factor influencing ampere-hour capacity is the temperature of the battery. Batteries are rated for performance at 20 or 25°C. Lower temperatures reduce usable capacity significantly. When designing the capacity of the system, the temperature of the battery in the coldest month of the year should be considered.

The nominal or rated capacity of battery (in Ah) is normally stated at a temperature of 20°C and a 10 hour rate of discharge or C/10 . It indicates that at this temperature, the battery will deliver the rated capacity C (usually in Ah) under a load I_{10} of C/10 A for 10 hours, when the cut-off voltage V_f will be reached. In most PV systems, the capacity of real interest is that at a much lower rate (100 hours or greater).

The state-of-charge (SOC) of a battery is defined as the ratio of remaining or residual capacity to the rated capacity. It is usually expressed as a percentage.

SOC = [Remaining capacity/rated capacity] × 100%.

The depth-of-discharge (DOD) is the ratio of the number of ampere-hours discharged to the rated capacity. It is the complement of the state-of-charge (SOC) (e.g. 25% DOD = 75% SOC).

When a battery undergoes a period of discharging and charging, it is said that the battery has made a cycle. Different PV applications may be characterized according to whether they typically cause shallow cycling (0–20% DOD), moderate cycling (20–50% DOD) or deep cycling (50–80% DOD) of the battery (Figure 6.3).

In a PV system, the daily DOD experienced may not be the same each day, owing to differences in availability of sunshine. In addition to daily cycles, the battery may be required to provide seasonal storage during winter or during the heavily overcast periods common in cold climates. It is also normal to allow for occasional deep cycling if something goes wrong with the system or if there is unexpectedly little sunshine.

This emergency reserve capacity is known as the *autonomy* of the PV system and is usually quoted as the number of days for which the average load can be powered without any PV charging at all, assuming that the battery is at 25°C and is initially fully charged. However, in cold climate applications, the usable portion of this autonomy reserve

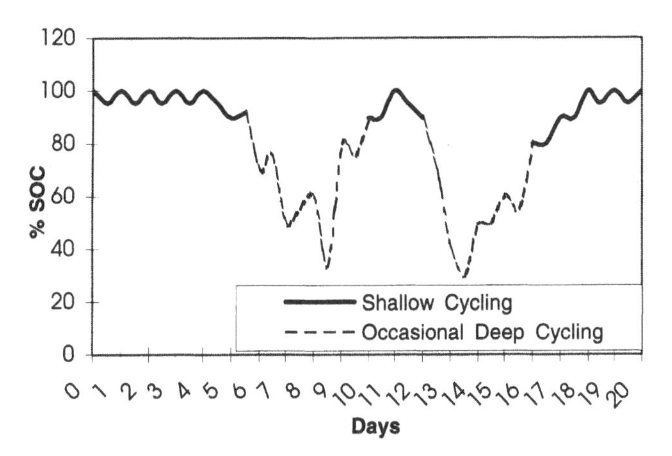

Figure 6.3. Deep and shallow cycling of batteries

(the 'true autonomy') is also influenced by the fact that lead–acid batteries are prone to freezing when discharged. We shall see in a following section that this true autonomy can be shortened dramatically because it may not be possible to deep discharge the battery at cold temperatures.

When a discharged battery is recharged at a suitable current, the voltage of the battery will rise until it reaches a point where some of the charging current will be consumed by *gassing*, which is the decomposition of some of the water in the electrolyte into hydrogen and oxygen. The battery can no longer be charged at this current without excessive water loss and the charging should be halted or modified by a 'charge controller' in order to limit this gassing. The voltage at which the charge current is halted or reduced is called the *cut-off voltage* (see Chapter 7).

Cycle life and service life

During discharge and recharge cycles, the chemical changes in the active materials of the electrodes usually cause them to change in volume. In time, this can lead to some active material becoming isolated from the current pathway, either because it physically drops off or because an insulating layer builds up around it. This latter phenomenon is called *sulphation*. The available capacity therefore starts to be reduced after many repeated cycles. The cycle life is a measure of how fast this happens. It is normally quoted as the number of discharge cycles to a specified DOD that a battery can deliver before its available capacity is reduced to a certain fraction (normally 80%) of the initial capacity.

The depth of discharge and its frequency can vary a great deal from one type of PV application to another. The following examples are in order of decreasing daily cycle depth:

- In PV–diesel hybrid systems used as remote area power supplies for homes and farmsteads, the batteries are cycled deeply, up to 80% every day in extreme cases.
- In summer cottages, PV systems are mostly used for lighting, TV and other household appliances, which may be directly powered by the battery or powered through an AC inverter. These systems are designed to be occasionally deep discharged. This may occur, for example, when the site is used only during weekend visits. The battery may be deeply discharged during a visit (up to 80%), but such a deep discharge would not occur every day.
- In small monitoring devices (including portable telemetry systems), the system autonomy is often only 5 to 10 days and the load may be used continuously. The average daily depth of discharge is usually designed to be around 15 to 30%, with a maximum of 80% DOD for extreme operation conditions.
- In remote telecommunications systems, customer-specified autonomies of 30 days and more are not uncommon. The load is generally continuous (i.e. on 24 hours per day), so only about half of it contributes to daily cycling. Under these circumstances, average daily cycling is between 2 and 4%.

In stand-by use or under small-load conditions, cycling of the battery may be very low and its lifetime will not be dictated by its cycle life, but by the *internal corrosion processes* that proceed continuously at a low rate. This depends heavily on the battery temperature and on the *float voltage* at which the battery is maintained when charged. Because the internal corrosion processes approximately double in rate for every 10°C rise in battery temperature, the float service life will approximately halve for every 10°C rise in temperature. A quoted *float service lifetime* is stated as a certain number of years, at a specified temperature and float voltage, before the available capacity drops to 80% of its initial value. For example, a certain tubular plate lead–acid battery may have a specification of 15 years at a float voltage of 2.25 V per cell at 20°C ambient temperature.

Factors affecting the lifetime of a battery

Factors affecting the lifetime of a battery in a PV system can be divided into three main categories:

- *primary factors*: cycle life and temperature, which are common to all applications of batteries.
- *secondary factors*: effects of periods at low SOC (e.g. sulphation in lead–acid batteries) and low overcharge, which are particular to PV use;
- *catastrophic factors*: manufacturing faults, undersizing, user abuse, freezing, etc., which are avoidable.

As long as the catastrophic failures are avoided and the correct choice of battery is made, the lifetime of the PV battery will be limited either by its cycle life or its internal corrosion processes, which are very closely related to its float life. While the cycle life is more or less independent of ambient temperature, the resistance to internal corrosion falls rapidly at higher ambient temperatures. Whether the cycle life or the temperature-dependent corrosion limits the battery life depends on the particular details of the PV system, the type of battery used, the daily depth of discharge and the average ambient temperature experienced.

We shall see in the following sections how different battery types are affected by these factors and how they can be minimized by good designs and an understanding of the characteristics of each battery.

LEAD–ACID BATTERIES

Lead–acid (Pb–acid) batteries have been in use for over 150 years and continue to dominate the automotive, stand-by and traction markets. Different types of cell construction are designed for specific applications and may not necessarily work well for other applications. Proper selection, therefore, requires knowledge of battery principles and available battery types.

Basic chemistry

In the charged state, the lead–acid cell has a positive plate with lead dioxide (PbO_2) as the active material and a negative plate with high surface area (spongy) lead active material. The plates themselves, which support the active materials and carry the electrical current, are made from lead metal, normally alloyed with traces of other elements. The electrolyte is a fairly concentrated sulphuric acid solution in water (about 400–480 g/litre). The concentration of acid is quite conveniently measured by its density. In a fully charged battery it is normally between 1.24 and 1.28 kg/litre. The specific gravity (SG) is a measure of the density relative to that of water. It is the same as the value in kg/litre, but without the units (i.e. 1.24–1.28). In some cases, the specific gravity is quoted as a number 1000 times the true value (e.g. 1240 instead of 1.24); hydrometers, the devices used to measure specific gravity, are often marked in this way in order to avoid using a decimal point, which may be difficult to see.

When current is drawn from the battery, the lead dioxide of the positive plate and the spongy lead of the negative plate are both converted to lead sulphate by reaction with the sulphuric acid. The reaction causes electrons to flow through the external circuit from one plate to another. Between the plates, the current through the electrolyte is carried by hydrogen ions (H^+). The reversible chemical reaction for the process is as follows:

$$PbO_2 + Pb + 2H_2SO_4 \leftrightarrow 2PbSO_4 + 2H_2O$$

charge discharge

As the battery discharges, the concentration of acid in the electrolyte decreases and the water content rises. The density of the electrolyte is thus closely related to the battery state of charge (see Table 6.1).

Table 6.1. Relationship between the specific gravity and the SOC of a typical lead–acid battery

Specific gravity @ 15°C	Battery state
1.270–1.285	Fully charged
1.190–1.210	Approximately half charged
1.120–1.150	Fully discharged

Batteries are recharged by a supply of direct current from an outside source, such as a PV array or a constant-current charger. The reversible reaction produces sulphuric acid at the two plates which increases the concentration of acid in the electrolyte. Both the voltage and the specific gravity of the battery will vary during charging and discharging (see Figure 6.4). The actual values depend on the state of charge, the current (or rate), the temperature and, especially at the end of charge, the type of battery.

In general, the voltage follows the pattern shown in the graph. Note that this voltage is different from the open-circuit voltage of the battery which is lower during charging and higher during discharging.

The open-circuit voltage, which indicates the state of charge of the battery better than the voltage under load, will typically vary from around 2.11 V/cell for a fully charged battery to 2.04 V/cell at 50% SOC to 1.98 V/cell for a fully discharged battery. However, if charging or discharging is interrupted in order to try to measure the open-circuit voltage, it can take many hours for the battery voltage to stabilize. This is because during charging and discharging the composition of the acid is not the same at all points in the battery, especially within the fine pores of the active material itself, and it takes some time for natural diffusion processes to equalize the acid composition.

During charging, sulphuric acid is generated at both plates. This tends not to mix very thoroughly with the more dilute acid between the plates and, because it is significantly denser, will tend to fall to the bottom of the cell. When the battery is recharged, the acid does not mix completely and on cycling there is a tendency for the acid at the top of the cell to become more dilute than the acid at the bottom. This phenomenon is called *stratification*. Only at the very end of

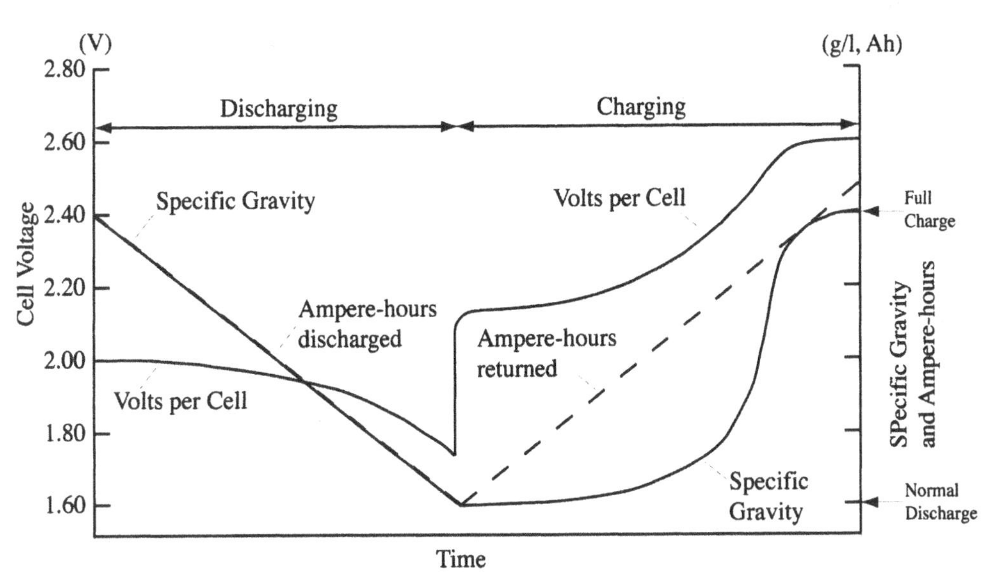

Figure 6.4. State of charge of a typical lead–acid battery[1]

charge, when gassing occurs, does the acid at the top of the cell receive some mixing with the acid between the plates.

Construction of lead–acid batteries

While we do not want to dwell on the different aspects of battery construction, it should be noted that different grid alloys and plate constructions will affect the performance of lead–acid batteries significantly.

The active material on the positive plate, lead dioxide, is a soft metal that usually is too fragile for the thin, open-mesh grids of pasted plates used in most batteries. Substances such as antimony and calcium are added in concentrations ranging from 0.1% to 11% of the grid weight to the positive plate to form a lead alloy that is stronger and easier to cast during manufacture than pure lead. Other metals such as tin, arsenic and silver may also be added to improve metallurgical, mechanical and casting properties, but antimony and calcium are the two alloys that are the most common for solar batteries. The negative plate does not usually require strengthening since its active material forms a strong bond with the grid.

Alloying 5% to 11% antimony with lead (high-antimony alloy) was the usual practice in the battery industry until about 15–20 years ago and such alloys are still used today for some applications. The main harmful effect is on the self-discharge rate, which increases dramatically as a battery with high-antimony plates ages. However, deep-cycling performance and cycle lifetime are markedly improved by the addition of antimony.

In order to minimize the effects on self-discharge and charge efficiency, while retaining the mechanical strength and the benefits of antimony on deep-cycling performance, *low-antimony* grid alloys have been developed. Here the antimony content is reduced to 1–3% and a third alloying element, usually selenium or arsenic, is added to obtain the required hardness. Traces of other elements are also added in many cases. Low-antimony alloys give a lower self-discharge rate than high-antimony alloys and the self-discharge rate does not change much even as a low-antimony battery ages.

The lower amounts of gassing (which results in water consumption) with low-antimony alloys are still too high to allow the use of cells that are fully or partially 'sealed' or 'maintenance-free', i.e. cells to which no water need be added over the lifetime of the battery. For this purpose, a different family of lead alloys (*lead–calcium*) has been developed, based on a small addition of calcium (typically 0.06% to 0.09%) to achieve the required hardness. While the self-discharge rate and charge efficiency of batteries with lead–calcium alloy plates are an improvement over those with low-antimony plates, the cycling and deep discharge recovery are markedly worse. To counteract this effect, tin is very often added to lead–calcium alloys. Batteries with such lead–calcium–tin alloys often have cycle lives that approach those for low-antimony plates.

The second main classification of lead–acid batteries is the construction of the positive plate. The positive grids of lead–acid cells are manufactured in three main forms for different applications. Negative plates in nearly all lead–acid batteries are of the flat (or pasted) type.

Flat-plate positive grids (Figure 6.5) are the most common type since they lend themselves readily to automation. The method of manufacture is to mould the grid as a flat lattice of rectangular holes. These holes are then filled with a paste of lead oxide, which is converted to active material at the

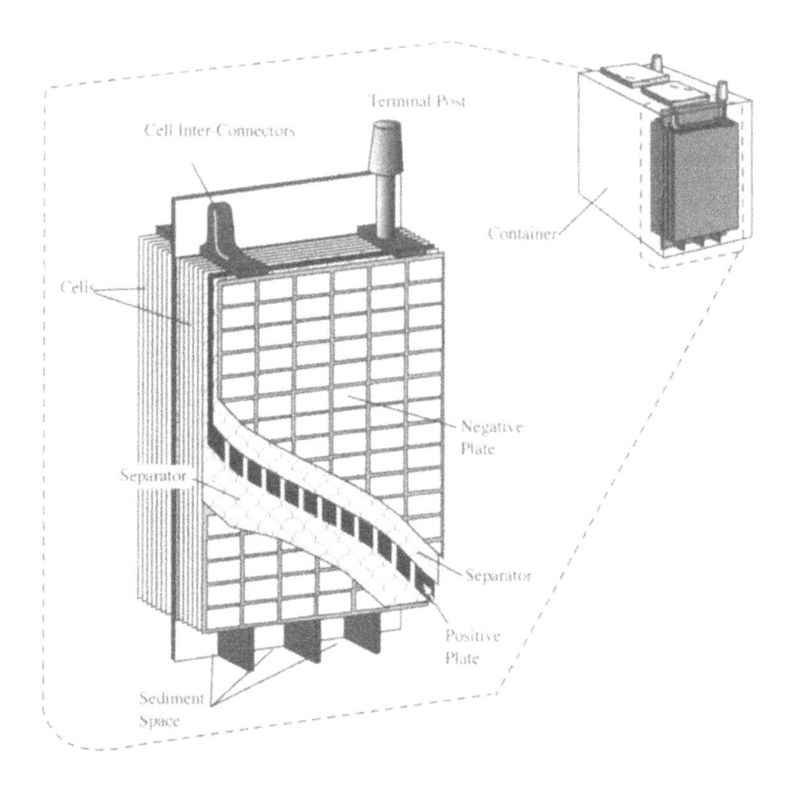

Figure 6.5. Details of a 12 volt flat-plate open lead–acid battery

forming stage (electrically charged into the required chemical composition).

The positive plates can also be made as a series of lead rods or tubes. These batteries are called tubular, multi-tubed, armoured or clad batteries. The active material is packed around each rod and held in place by strong tubes of insulation material. The tubes reduce shedding of active material from the electrode and thus increase their deep cycling capabilities. They are also more resistant to vibration.

The positive plates of Planté batteries consist of solid sheets of lead 7–12 mm thick with grooved faces to increase surface area. The surface is coated with paste and formed to produce a coating of porous active material. The thickness of lead acts as a reserve of active material. When active material falls off, the lead that becomes exposed changes to lead dioxide during charging.

In PV batteries, only flat-plate and tubular-plate positives need be considered. Planté's are usually not suitable because of high corrosion when being charged near their open-circuit voltage.

A last distinction between different types of lead–acid battery is whether they are open (vented) or 'sealed' (valve-regulated). In open batteries, overcharge results in the conversion of water into hydrogen and oxygen gases, which are lost to the atmosphere. Water has to be added to these batteries from time to time to make up this loss. In valve-regulated batteries, overcharge results in oxygen gas production at the positive plate, but because the space between the plates is not completely filled with acid, the oxygen gas can reach the negative plate, where it reacts with hydrogen to form water. This recombination of oxygen gas can only proceed at a certain rate. If the charging current is too high, then oxygen gas pressure will build up inside the cell and eventually the safety valve will release oxygen (and some acid spray) into the atmosphere. This will result in permanent loss of water.

Types of lead–acid batteries by intended application

The following applications describe the main battery types:

Car and truck (SLI)
SLI (starting, lighting, ignition) batteries are manufactured for the automotive industry. The main requirement of this battery is that it can deliver a very large current for a very short period of time. It is usually discharged by only a small fraction of its capacity since, once the engine is running, the battery is recharged by the car's alternator. High currents call for thin plates to provide a large surface area of active material to be in contact with the electrolyte. As a consequence, both the cycle life and the resistance to plate corrosion are correspondingly low.

Car batteries are not suitable for photovoltaic applications since they cannot withstand many deep discharge cycles. They are being used in many developing countries, however, because more suitable batteries are not available or are too expensive. A low-antimony car battery that undergoes shallow cycling (10% maximum depth-of-discharge) in a PV system can have a lifetime approaching that which it would have in a car (3 to 5 years at 20°C or 2 to 2.5 years at 30°C). Truck and bus batteries of the 'heavy-duty' type have thicker plates and can withstand more deep-discharge cycles.

'Maintenance-free' car batteries are made with lead–calcium positive alloys. The rate of gassing of these batteries is minimal and under normal operation they do not require that their electrolyte be 'topped up'. They are often called 'sealed', meaning that the user cannot refill them, but they are not truly sealed batteries – they have a pressure vent. Care must be taken to limit end-of-charge voltage to under 2.35–2.4 V per cell. At higher voltages, the gassing reaction will be too fast and venting of gases will occur. These batteries can also suffer from irreversible sulphation if deeply discharged and not recharged immediately.

Marine/leisure/solar

Marine/leisure batteries are usually made with a low antimony (0.5–3%) positive grid alloy and medium-thickness plates. They have a high cycle life for deep discharge applications. However, they self-discharge quickly and have high rates of gassing at the end of charge. These batteries cannot be sealed and must be maintained regularly by adding distilled water.

Special flat-plate batteries carrying a 'Solar' label are available in some countries for use in consumer PV systems. These are usually an adaptation of a battery manufacturer's truck or marine battery. They will normally have similar positive plate thickness to a truck battery as well as an improved separator between the plates and a low antimony positive grid alloy.

Semi-traction/golf cart

These batteries are used to power small specialized electric vehicles such as wheelchairs and golf carts, where cycling can be very important. In practice, they may be little different from leisure batteries, although a higher-antimony positive grid alloy is more common. Low-antimony semi-traction batteries are suitable for domestic PV systems.

Golf-cart batteries generally have a high energy density per unit of mass (Wh/kg). This means that they have thinner grids and less electrolyte than other semi-traction batteries. They will consequently not tolerate as many deep discharges and may be less suitable than marine, solar or semi-traction batteries.

Traction

Traction batteries are used to provide power to fork-lifts and other low-speed electric vehicles. These batteries can be deep-discharged about every 24 hours and their cycle life is typically 1200 to 1500 cycles. The plates are large and thick and the positive plate (flat or tubular) will contain a high level (4–8%) of antimony to improve its mechanical strength. Unfortunately, this increases self-discharge. While this is not a problem in traction applications where the batteries are recharged daily, it makes them a poor choice for solar applications.

Stationary

Stand-by batteries provide emergency power to critical equipment such as telephone lines, railway lighting and emergency systems; they are also common in uninterruptible power supplies (UPS) for computers. These batteries are usually float charged (i.e. kept fully charged at all times by a very small current) and power the load whenever there is a blackout. Stand-by batteries need to be very reliable, have a low rate of self-discharge and a long calendar life of ten years or more. They usually have thick, heavy plates with heavy connections. The positive plate grids are usually vertical tubular rods. They are not normally designed for deep cycling, although, as a result of their robust construction, they will survive a certain number of deep discharges.

Valve-regulated 'sealed' batteries

Fully sealed lead–acid batteries are used whenever regular maintenance is not practical. They are designed in such a way as to eliminate the loss of water under normal operating conditions. The capacity of the positive plate is slightly less than that of the negative plate. As a result, the end-of-charge gassing reaction begins at the positive plate sooner than it does at the negative plate. Oxygen gas produced at the positive plate migrates to the negative plate, where it reacts with lead and sulphuric acid to produce water and lead sulphate. Further charging changes this lead sulphate to lead and restores the chemical balance of the cell. The cycle only works with oxygen; hydrogen gas recombines very slowly and thus the battery cannot sustain high rates of charge without some loss of hydrogen gas.

The above recombination reaction cannot occur in a flooded battery because the liquid electrolyte prevents the oxygen gas from reaching the negative plate. In valve-regulated batteries two approaches are used to eliminate the liquid electrolyte and thereby circumvent this. One approach uses a spongy separator material called an absorbent glass mat (or AGM). There is little free electrolyte in the cell – it is electrolyte 'starved' – and oxygen can easily pass through the porous separator. The other approach uses an electrolyte made into a jelly consistency (or 'gel') using silica-based additives. Microscopic cracks in the gelled electrolyte permit oxygen to pass through the electrolyte.

Sealed batteries have a compact structure because the plates are thin and held in close contact with the separators. Active material is prevented from falling off the plates by the gel-like electrolyte or AGM separator. Good support of the plates permit the cell to survive a complete discharge without permanent damage. Their life cycle is short if used continuously for deep discharge applications, however.

Sealed batteries are very suitable for photovoltaic applications. They can be placed in any position and do not require any maintenance. Their main disadvantages are high cost and short cycle life when deeply discharged. Though they have a low self-discharge rate, they can be permanently damaged by sulphation if stored for several months without charging.

Because there is little free electrolyte in AGM batteries, they are more freeze tolerant than gel batteries.

Effect of low temperature on lead–acid batteries

Lead–acid batteries are affected by low temperature in several ways.

Effect of temperature on lifetime

Because the internal corrosion processes in a lead–acid battery approximately double in rate for every 10°C rise in battery temperature, the service life will approximately halve for every 10°C rise in temperature above the stated value. Conversely, this service life may double for every 10°C decrease. However, below +10°C, other factors ultimately reduce the lifetime of the battery and its service life will not increase much.

Effect of temperature on capacity

We have seen that the battery capacity will decrease significantly with a decrease in temperature for all batteries (see Figure 6.2). This also shows that at higher discharge rates, the capacity of the battery will be further reduced. For example, if the danger of freezing is ignored, at -20°C battery capacity will be reduced by 40% at a *C/5* rate but only 25% at a *C/100* rate. Battery data sheets may have more accurate information about the effect of temperature on capacity for a particular brand of battery.

Effect of temperature on end-of-charge

The charging characteristics of a lead–acid battery will also be affected by temperature. The colder the battery, the less charge it will accept at a given voltage. In order to be fully charged, the cut-off voltage will need to be set higher in cold temperatures. Figure 6.6 shows the effect of different temperatures on the recommended end-of-charge voltage for a typical vented lead–acid battery. Even with temperature compensation, it can be difficult to fully charge a cold battery, especially at higher currents.

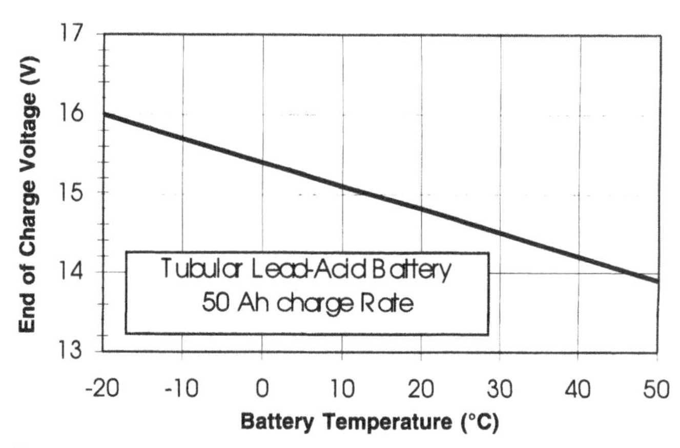

Figure 6.6. End-of-charge voltage at various temperatures for a typical 12 V vented lead–acid battery

Freezing of electrolyte

More significantly, as a lead–acid battery is discharged, the water content of the electrolyte increases and the acid concentration decreases. In a fully discharged battery, the acid concentration will fall to around 120–230 g/litre (SG 1.08–1.14), depending on the rate and the amount of free acid. As the acid concentration falls, the freezing point rises from between -45 and -66°C in a fully charged battery to between -6 and -13°C in a fully discharged battery. Thus, there is a danger that the electrolyte of a discharged lead–acid battery may freeze at low temperatures. A battery with frozen electrolyte cannot be further discharged and recharge is difficult. Most open lead–acid batteries will suffer permanent damage if frozen.

All types of lead–acid batteries will run the risk of freezing if they are 100% discharged at the 100 hour rate at temperatures below about -10°C. If 80% discharged at the 100 hour rate, sealed lead–acid batteries risk freezing below about -12°C and vented lead–acid batteries risk freezing below -15 to -20°C, depending on the exact type. Lead–acid batteries of any type that have been 100% discharged at the 100 hour rate at 20°C or above (where they can give their full capacity) should not then be subjected to temperatures below 0°C (see Figure 6.7).

One way to prevent freezing in a battery that is to be used in a cold climate is to increase the initial electrolyte concentration (i.e. specific gravity). This will enhance the performance of the battery under extreme conditions while giving better freeze protection to the battery. However, it will accelerate the internal corrosion processes and may thus reduce the lifetime of the battery, especially if the battery is also subject to higher-temperature periods.

Effect of stratification and temperature

Stratification (see above) increases the risk of electrolyte freezing, because the weaker acid at the top has a higher freezing point. It is difficult to be precise about the damaging

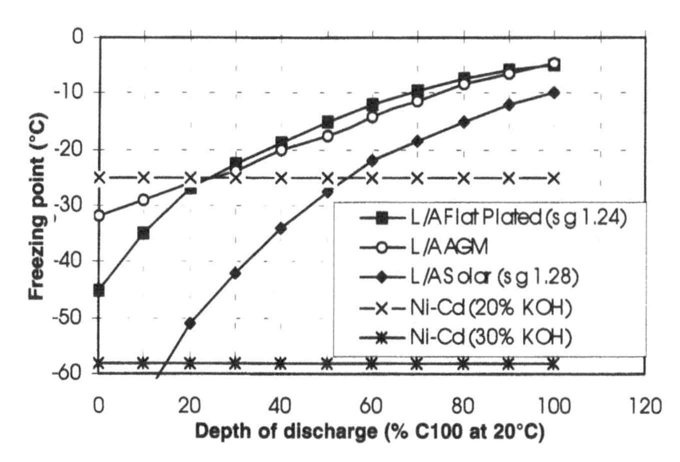

Figure 6.7. The freezing point of various lead–acid (L/A) and nickel-cadmium (Ni-Cd) batteries as a function of SOC[2]

effects of localized freezing, but it is wise to try to avoid it by not allowing the battery to be deep discharged to any great extent if temperatures are going to be below –5°C. Furthermore, a stratified battery has highly acidic electrolyte at the bottom of the plate, accelerating corrosion and reducing lifetime.[3]

Effect of temperature on measuring state of charge

Temperature affects both the specific gravity and the open-circuit voltage of a battery. Since these parameters are normally used to measure the state of charge of the battery, their precision can be an important factor in proper battery maintenance. This can be particularly significant with batteries that are not used for long periods, but which must be kept fully charged so that they are not damaged by the cold.

The specific gravity of an electrolytic solution is usually given at a temperature of 15°C. At a lower temperature, its reading will be noticeably higher, suggesting a state of charge that is higher than in reality. For example, electrolyte with a specific gravity of 1.25 at 15°C will have a specific gravity of 1.27 at –15°C. Appendix C shows how to calculate the specific gravity at 15°C for a measurement at a temperature other than 15°C.

We have also seen that stratification affects the specific gravity reading of the state of charge since the concentration of acid will not be uniform within the battery. A reading at the top of the battery would detect that the battery is less charged than a reading at the bottom of the battery.

The open-circuit voltage will also vary slightly according to the temperature of the battery. This temperature correction is uneven. At a very low state of charge, the temperature coefficient will be almost nil, reaching a peak of 0.25 mV per °C deviation from 25°C at a specific gravity of 1.15 and dropping to 0.2 mV/°C at a specific gravity of 1.28 or full charge.

Effects of deep discharge on lead–acid batteries

A lead–acid battery that is left discharged for a long period of time may suffer from a permanent loss of capacity due to sulphation. This occurs when the small lead sulphate crystals formed during discharge are transformed into larger crystals, which are not easily dissolved on the subsequent recharge. In addition, a thin layer of an electrically insulating type of lead sulphate can form on the grid material during a very deep discharge, reducing battery efficiency. Batteries designed for deep discharge use lead alloys that minimize this latter problem.

Another effect which may occur after a very deep discharge is that the acid becomes so dilute that the lead sulphate dissolves somewhat and, on recharge, fine needles (or dendrites) of metallic lead are plated back on the negative plate. These can cause short circuits within the battery.

In cold-climate applications, we have seen that another important effect of deep discharge is that the battery will freeze at a higher temperature as the electrolyte becomes more dilute.

For all these reasons, it is common practice to ensure that a lead–acid battery is not discharged beyond 80% DOD, even when the battery is designed for deep cycling. This is done by using a device that disconnects the battery from the load when the voltage has fallen to a level roughly corresponding to the desired maximum depth-of-discharge.

NICKEL–CADMIUM BATTERIES

Nickel–cadmium (Ni–Cad or Ni–Cd) batteries are sometimes used instead of lead–acid batteries in applications that require:

- a longer operating life in either cyclic or stand-by use;
- a higher energy to weight ratio (portable / mobile uses);
- operation at very low (sub-zero) or very high (over 40°C) temperatures;
- very low capacities (less than 1 Ah).

In PV systems, their selection in preference to lead–acid batteries is restricted to similar criteria, largely because they are typically three to four times more expensive per unit of energy storage capacity than most types of lead–acid batteries.

Basic chemistry

In the charged state, the nickel–cadmium cell has a positive plate with nickel oxy-hydroxide (NiOOH) as the active material and a negative plate with finely divided cadmium metal. The plates themselves, which support the active materials and carry the electrical current, are made from nickel or nickel-plated steel. The electrolyte is a fairly concentrated solution of potassium hydroxide (KOH) in water (20–35% by weight). Lithium hydroxide (LiOH) is added to the electrolyte of some cells to improve cycling performance.

When current is drawn from the battery, the NiOOH of the positive plate is reduced to $Ni(OH)_2$ and the cadmium metal of the negative plate is oxidized to $Cd(OH)_2$. The reaction causes electrons to flow through the external circuit from one plate to another. Between the plates, the current through the electrolyte is carried by hydroxyl ions (OH^-). The reversible chemical reaction for the process is as follows:

$$2NiOOH + Cd + 2H_2O \underset{\text{charge}}{\overset{}{\leftrightarrow}} 2Ni(OH)_2 + Cd(OH)_2 \quad \text{discharge}$$

The charging process at the positive (nickel) electrode is not very efficient. Towards the end of charging, the recharge reaction must compete with a reaction that evolves oxygen gas at the surface of NiOOH. This effect is more pronounced in thicker electrodes. In general, oxygen gas production starts around 80% SOC and increases as the SOC approaches 100%.

In contrast, the charging of the cadmium (negative) electrode is quite efficient. Hydrogen gas is only produced in appreciable amounts (in an open cell) when all the cadmium hydroxide has been converted to cadmium metal. In open cells, this point is marked by a sudden increase in cell voltage, which is a convenient indication that the battery is fully charged. In 'sealed' cells, where hydrogen evolution at the negative plate is deliberately suppressed, the attainment of full charge is not marked by any sharp increase in voltage at all, making it impossible to use a conventional voltage regulator to control the charging.

Types of nickel–cadmium batteries

Nickel–cadmium batteries are available in two basic types of construction, called 'sintered' and 'pocket plate'. Vented nickel–cadmium cells can be made from either sintered or pocket plates while small sealed cells generally use sintered plates.

Sintered plates are made of nickel powder pressed at high temperature, in a process called sintering, to form plates that are strong and porous like a sponge. The voids in the sintered plates are then filled with active materials to form the positive and negative plates of the cell. This construction permits lower internal resistance and sensitivity to variable temperature operation.

Sintered Ni–Cad batteries are available in the same sizes as zinc dry cells (i.e. common non-rechargeable torch batteries). Although the voltage of nickel–cadmium (1.25 V) is less than that of the zinc dry cells (1.5 V), they can be used interchangeably in many applications. Sintered batteries are also available in 12 V formats and in capacities ranging from 0.1 to 25 Ah.

Most sintered cells are sealed so that the electrolyte cannot leak out and the battery requires no topping up. However, since gassing does occur, the cells are constructed in such a way that the negative plate is larger, causing the oxygen to be produced first by the positive plate. This oxygen is allowed to pass freely from the positive to the negative plate, where it is absorbed forming cadmium hydroxide. Further overcharging decomposes the cadmium hydroxide to cadmium and water to complete the cycle. Sealed cells can be overcharged continuously without loss of gasses provided that the current is not too high.

Sealed sintered cells should only be charged at constant current. The recommended charging current is usually the C/10 rate and the battery requires 14 to 16 hours to recharge. Surprisingly, the charging efficiency for sealed cells is worse when charging more slowly than the recommended C/10 rate and in some cells there is no charging at currents below the C/40 rate. This means that it is better to charge sealed cells at a constant current with a regulator rather than directly from solar modules which supply a varying current.

Sintered plate Ni-Cads can suffer from a '*memory effect*' phenomenon during discharge, which can affect the discharge voltage. This effect is caused by repeated partial discharge to a particular depth. When subsequently discharged beyond that level, the battery will operate at a much diminished voltage. The degree to which the voltage is reduced depends on the number of times the battery has been cycled to that level. The temporary effect can be erased by subjecting the cell to a complete discharge/charge cycle.

In a *pocket-plate Ni–Cad battery*, the active materials are held on the electrodes in pockets and the pocketed rectangular plate is suspended in the electrolyte. The plates are rigid and remain flat during cycling of the cell so that the separators need only be plastic rods or an open mesh that sets the spacing between the plates and the container.

Pocket-plate cells are generally vented. Hydrogen and oxygen produced by gassing must be replaced occasionally with distilled water. The gassing vent is not a simple hole but a one-way valve, preventing carbon dioxide from the air being absorbed by the electrolyte. Pocket cells usually have higher capacities than sintered cells (from 5 Ah to 1300 Ah), extended cycle life and the ability to withstand long periods at partial states of charge without sustaining damage.

Characteristics of nickel–cadmium batteries

There are several differences in Ni–Cd characteristics compared to those of lead–acid batteries:

- The potassium hydroxide electrolyte is not consumed in either the overall charge or discharge reactions, which means that:
  - The electrolyte density (or specific gravity) does not change as the battery is charged and discharged and thus the measurement of electrolyte density is no measure of the state of charge and there is no problem with stratification.
  - The freezing point does not vary with the state of charge. It does, however, depend on the strength of the KOH electrolyte, varying from –25°C for 20% KOH (density 1.19 g/litre) to –58°C for 30% KOH (density 1.29 g/litre). At concentrations above 32% KOH (density 1.31 g/litre, freezing point –61°C), the freezing point rises (e.g. to –50°C at 35% KOH (density 1.34 g/litre).
  - Capacity does not increase very much at very low discharge rates, even if there is a large reserve of electrolyte.
- There is no damage if a Ni–Cd battery is left in a low state of charge. In fact, batteries are often transported or stored in this condition, to reduce the risk of short circuits.
- Internal corrosion does not occur to any great extent below 35 or 40°C. This is quite different from the case of lead–acid batteries, where an increase of 10°C (e.g. from 20° to 30°C) can halve the life if cycling is not the main life-determining factor. The lifetime of Ni–Cd batteries on float duty is typically reduced by 20% for every 9°C that the battery temperature exceeds 25°C.
- Nickel–cadmium batteries have a lower charge efficiency than lead–acid batteries.
- Nickel–cadmium batteries have a lower self discharge rate than lead–acid batteries.
- Open-circuit voltage is more or less independent of the state of charge. It is around 1.25–1.3 V/cell at room temperature. On charge and discharge, the voltage (especially at low rates) is generally more constant than for lead–acid batteries, except at the beginning and end of the process.

BATTERIES FOR COLD-CLIMATE APPLICATIONS

While the main criterion for selecting a battery is usually cost, this costing must be carried over the lifetime of the battery. Other factors that will have an impact on cost, such as transportation, maintenance and disposal, must also be evaluated when making the choice for the proper battery. The following guidelines on battery selection for cold-climate applications are given as indications only.

Worst-month average battery temperature is below –10°C

This section only applies if the battery is going to be less than fully charged in such cold conditions. For systems where electricity is used only during the summer months and the battery is left charged at the beginning of winter, see the next section.

Note that the battery capacity needs to be corrected for low temperature (see Figures 6.2 and Figure 6.7). Use the minimum temperature expected for months when the battery will be used.

Nickel–cadmium (open, pocket-plate or fibre-plate) is the best option to avoid any risk of battery freezing. If operation of the battery will be below –20°C, check that the electrolyte is approximately 30% KOH (density around 1.29 g/litre) and not 20% KOH (density around 1.19 g/litre).

The alternative to using nickel–cadmium is to oversize a lead–acid battery by adjusting the minimum depth of discharge to a suitably high value so that the acid will never reach a density where it can freeze (see the earlier section on the effects of temperature). High-density acid should be specified for open lead–acid batteries at low temperatures. The technique still carries some risks of freezing if the battery is overdischarged and battery sizes may become quite large for sites where the minimum battery temperature will be below –20°C. If sealed lead–acid batteries are required, use the AGM type, not gel, to avoid damage by freezing.

Another alternative, if feasible, is to consider providing heat to the battery enclosure (or the battery itself) so that its minimum temperature is 0°C or above. Then the guidelines in the next section can be followed instead.

Worst-month average battery temperature is above –10°C and below 20°C

Note that the battery capacity needs to be corrected for low temperature (see Figure 6.2). Use the minimum temperature expected for months when the battery will be used.

Average daily cycling in such climates is likely to be shallow (< 10%), because of the need for large seasonal or autonomy reserve. If, however, cycling will be deeper than this, the lifetime of batteries designed to give fewer than 200 cycles to 80% depth-of-discharge could be shortened.

The different options are:

- *Required life around 10 years.* Most cost-effective are low-antimony open tubular stationary or thick flat-plate industrial lead–acid types.

- *Required life around 20 years.* Open pocket-plate or fibre-plate nickel–cadmium batteries will give this life, but replacing one of the above open lead–acid types of battery after ten years is often more cost-effective (depending on the actual capacity required and thus the relative price per kWh).
- *Sealed batteries.* If sealed lead–acid batteries are required and the lowest temperature will be 0° or below, use the AGM type to avoid damage due to freezing. Above 300 Ah, heavy-duty AGM types will give about 15 years service. Below 300 Ah, medium-duty AGM types will give about seven years service and will normally be more cost-effective, even if replaced once.
- *Low-cost options.* Flat-plate open lead–acid types: solar, truck starter, marine, leisure or semi-traction batteries (in approximate order of preference). Low-antimony types are strongly recommended and service life should be around five years or more.

PV BATTERY PROTECTION

Enclosure

All batteries should be protected from the outside environment by a suitable enclosure. This enclosure may be anything from a suitable box to a dedicated room or shelter. It must be made of material that is resistant to corrosion by the electrolyte. In particular, nearly all unpainted metals and acid-sensitive plastics, such as nylon, should be avoided near lead–acid batteries and metals such as aluminium and zinc (including galvanized coatings) should be avoided near Ni–Cd batteries.

Temperature protection

Low temperatures reduce available capacity and can also introduce the risk of freezing a partly discharged lead–acid battery. Battery banks must then be sized on the basis of the lowest battery capacity anticipated. This may result in a large battery bank. By increasing the temperature of the battery in winter, an enclosure can effectively increase the usable capacity of the battery bank and permit a smaller battery bank to be used.

In order to do this the following steps should be taken:

- Use a highly insulated enclosure to keep the heat produced by the battery inside as much as possible. All battery types will generate heat during operation, especially during overcharge. Note that proper ventilation space must be left between batteries so that the ones in the middle of stacked batteries do not overheat. All cells in a battery should operate as near as possible to the same temperature and therefore should be in the same thermal environment. This is especially important for valve-regulated batteries. For all battery types, avoid sitting individual blocks or cells near any heat source or near cold walls. Verify that the batteries are not so well insulated that heat generated by the batteries will elevate summertime battery temperatures above 30°C.
- Insulate battery enclosures and use water both as a thermal mass and a phase-change material. A study done at CEDRL[4] shows that water acts as a very effective phase-change material and can keep the battery at an acceptable operating temperature year round. System autonomy may be 80% higher than that provided by a normal insulated enclosure. By using phase-change materials, extreme temperatures in summer may also be alleviated.
- Use any available heat source (e.g. waste heat from a back-up generator) and even consider using passive solar techniques like allowing a window that transmits winter sunlight into the enclosure (but not higher-altitude summer sunlight).
- Consider heating the enclosure with electric resistance heaters powered by PV. For large battery banks it may be more cost-effective to size the PV power system so that the temperature of the enclosure is not allowed to go below a certain temperature. This reduces the need for large batteries and improves battery and system reliability.
- During charging, batteries will produce heat due to internal losses. Individual batteries in a battery bank should be kept at close to the same temperature to minimize disparities in operation. As a general guide, a minimum air gap of 5–30 mm (varies with battery size) should be present between the cases of neighbouring cells or battery blocks and racks or stands should allow free vertical movement of air between the cases. This requirement can be relaxed for batteries that will be charged and discharged very slowly.

Ventilation

During charging, open batteries will produce hydrogen and oxygen gases. Hydrogen is the lightest (least dense) gas of all and can temporarily concentrate at the top of the battery enclosure before diffusing throughout the free space of the enclosure. If the concentration of hydrogen in air exceeds 4%, there is an explosion hazard. Ventilation of open batteries should be designed to prevent this happening.

Sealed (valve-regulated) lead–acid batteries also produce small amounts of hydrogen due to internal corrosion processes. They also require some ventilation, although the amount required is more dependent on cooling needs than the removal of hydrogen.

The general conclusion is that no batteries should be put into a totally sealed enclosure.

Some batteries for domestic PV installations have a special cover that allows the gas generated to be led outside through a small tube. If this is used, make sure that the tubing is not bent through a sharp angle and that the end of the tube cannot become blocked.

REFERENCES

1. Photovoltaic Design Assistance Center (1991). *Maintenance and Operation of Stand-Alone Photovoltaic Systems.* Albuquerque, NM: Sandia National Laboratories.
2. Spiers DJ, Royer J (1998). *Guidelines for the Use of Batteries in Photovoltaic Systems.* Helsinki: Neste Advanced Power Systems; Varennes, Quebec: Natural Resources Canada.
3. Sauer DU (1997). Modelling of local conditions in flooded lead–acid batteries in PV systems, *Journal of Power Sources*, 64, pp.181–187.
4. Ross MMD (1998). Estimating wintertime battery temperature in stand-alone photovoltaic systems with insulated battery enclosures, *Renewable Energy Technologies in Cold Climates '98, Montréal, Canada, 4–6 May 1998.* Solar Energy Society of Canada, Inc., Ottawa, Ontario, Canada.

FURTHER READING

Barak M (Ed.) (1980). *Electrochemical Power Sources: Primary and Secondary Batteries.* Stevenage, UK: Peter Peregrinus.

Bechtel National, Inc. (1979). *Handbook for Battery Energy Storage in Photovoltaic Power Systems*, Report SAND80-7022. San Francisco, CA: Bechtel National, Inc.

Chang TG, Valeriote EM, Gardner CL (1990). A Study of the low-temperature charge acceptance of 500 Ah photovoltaic batteries: Laboratory and field tests, *Journal of Power Sources*, 32, pp.151–163.

Hill M, McCarthy S (1992). *PV Battery Handbook.* Cork, Ireland: Hyperion Energy Systems Ltd.

Ikkala O and Nieminen A (1990). Lead/acid batteries in Arctic photovoltaic systems, *Journal of Power Sources*, 31, pp.321–327.

Linden D (Ed.) (1984). *Handbook of Batteries and Fuel Cells.* New York: McGraw-Hill.

Sharpe TF, Conell RS (1987). Low-temperature charging behaviour of lead–acid cells, *Journal of Applied Electrochemistry*, 17, pp.789-799.

Spiers DJ, Rasinkoski AA (1996). Limits to battery lifetime in photovoltaic applications, *Solar Energy*, 58, pp.147–154.

7 Energy management issues

Raye Thomas and Jimmy Royer
(NewSun Technologies Ltd and Solener Inc.)

Electric loads in remote, cold-climate areas have traditionally been powered by diesel and gas generators or primary batteries. However, increasingly renewable-energy technologies such as photovoltaics or wind turbines are being used to supply part or all of the energy needs of the loads. PV is particularly attractive since it requires little maintenance and no exterior energy source. However, because of the variable nature of the solar resource, a PV array, unlike a genset, cannot produce power on demand. The energy produced by stand-alone PV systems usually needs to be stored to ensure that power is available to be called upon when needed – this complicates and constrains the energy management strategy.

NOTIONS OF ENERGY MANAGEMENT

As discussed in Chapter 2, three types of PV system can be used to power remote loads:

- the autonomous PV/battery system;
- the PV hybrid system;
- the PV system that is connected to a remote grid.

The first two types of system involve some kind of energy storage component which is usually based on an electrochemical battery. Such batteries have peculiarities that are often exacerbated when they are used in a cold-climate environment. Thus, the energy management strategy is concerned with properly managing the charging (either by PV alone or PV in conjunction with an auxiliary power such as a genset) and discharging of the battery, to ensure that load requirements are reliably met and battery lifetime is prolonged.

In the first two types of system, different energy management strategies are required depending on which of three levels of energy storage are required by a particular installation (see Figure 7.1):

- The first level requires that the energy being produced by the PV array be stored so that it can provide the energy demand of the load when it is required on a daily basis.
- The second level requires that the energy being produced by the PV array be stored to provide the energy demand of the load for a few days, when the sun is not available.
- The third level requires that the energy being produced by the PV array is stored for long periods varying from a month to four months (e.g. year-round requirements north of the Arctic Circle) in order to provide for seasonal fluctuation of insolation.

In the third type of system, one or many PV systems are connected to the grid to reduce the demands on the principal power system – usually a diesel genset. Energy management in this context includes managing the system to reduce fuel consumption, start-up frequency and runtime of the gensets.

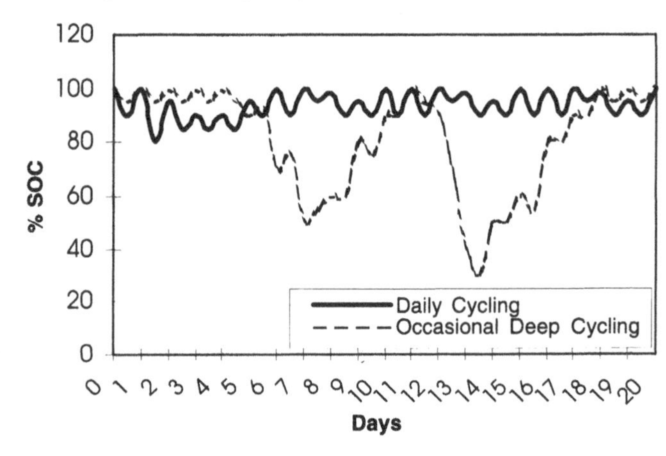

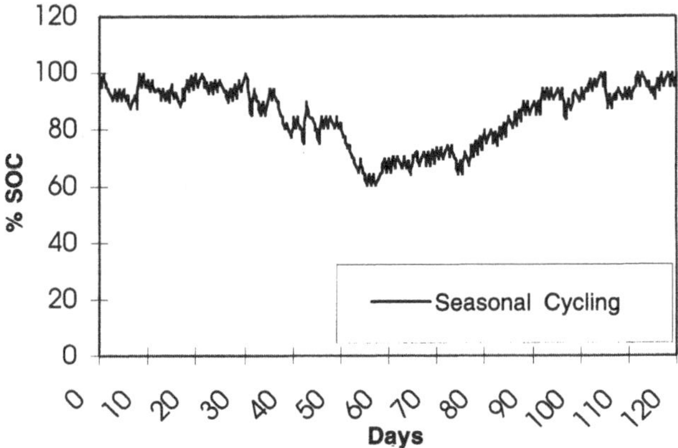

Figure 7.1. Typical battery state of charge for different charging cycles

Energy management strategies are defined by the very nature of the PV system itself. Thus energy management starts at the system design stage where the choice is made of the appropriate energy system for the load and site environment.

As stated, the first step in designing a proper energy management strategy is to choose the right kind of system and the right components for the proposed application. Earlier chapters have shown that photovoltaic systems can be used to power a wide range of loads in cold climate regions. Chapters 8 and 9 describe and show how to properly design stand-alone PV-only and PV-hybrid systems. For now, it is sufficient to mention that to achieve a high degree of reliability, a PV system must be able to provide power during the 'worst case' situations which are usually periods of adverse weather, seasonally low solar insolation values and unpredicted increased power demand.

OPTIMIZING BATTERY OPERATION

Optimization of the PV system involves understanding how each of the major subsystems affects the overall behaviour of the system. Since a stand-alone PV system is usually battery-based, the battery forms the core of system behaviour and any optimization will maximize the efficiency of the battery while protecting it against deep discharge and overcharging.

Controlling a PV battery

In order to optimize the charging process and minimize gassing of a battery, the usual practice in non-PV applications is to charge the lead–acid batteries using the IU method (fixed current until a certain voltage is reached, followed by a constant voltage period with decreasing current for a limited time). Nickel–cadmium batteries are normally charged with constant current for a fixed time.

Such methods are not possible in PV systems without wasting a lot of the energy available from the PV array. In most cases, the full PV current (which is variable) is used to charge the battery to a certain level, at which point an electronic charge controller either cuts off the PV charging current completely or reduces it to a lower level to finish the charging process. This process (called charge regulation) prevents excessive overcharge, which leads to high water losses and reduction of battery life.

Almost all charge controllers perform this charge regulation function and they are sometimes referred to as charge regulators. In addition, charge controllers may include other functions, such as:

- low-voltage load disconnect, to prevent over-discharge;
- display/monitoring features, to show battery voltage (state of charge), currents, etc.;
- remote alarm relays, to transmit signals indicating fault conditions (over- and undervoltage, array failure, etc.);
- temperature compensation;
- remote communications (in larger controllers).

Common types of charge regulators

Unlike some other power generators, PV arrays are generally tolerant of both short-circuit and open-circuit conditions. The array charging current may therefore be interrupted either by open-circuiting the array (series control) or diverting the array current through a low resistance path (shunt control), which effectively short-circuits the array. While this difference in controller type is important in terms of controller construction, the number of possible set points and the rate at which they can operate is more significant.

Most charge controllers use the battery voltage as a signal that the current must be cut or reduced. The voltage at which they do this is called the charge set point (or simply set point). Some of the larger microprocessor controllers use ampere-hour counting instead to estimate the actual SOC of the battery. This is especially true for sealed Ni–Cd batteries because voltage control is not possible as there is no sudden rise in voltage towards the end of charge.

In a simple one-step controller, when the battery voltage has risen to the charge set point, it stops the array from charging the battery. Charging is resumed only when the battery voltage has fallen to a much lower value (reconnect point). This reconnect point is normally set so far below the charge set point that the controller will not normally reconnect the PV array for some hours under average charging conditions, unless a heavy load discharges the battery during the day. This hysteresis (difference between charge set point and reconnect points) is necessary to prevent excessive switching of the array.

Larger series relay controllers can include more than one relay, each one connected to an independent PV sub-array. Each relay can have a different set point, so that they operate in sequence. The PV charging current is then reduced in a series of steps rather than cut all at once. The effect is often that the battery voltage stays almost constant after some of the relays have opened.

Solid-state switching controllers (shunt or series) can operate at very fast rates (even thousands of times per second). When the charge set point is reached, they can pulse the PV current on and off in order to keep the battery voltage as constant as possible. The on–off duty cycle will vary according to what current is available from the array and the actual SOC of the battery. Such control is sometimes referred to as switching, pulse-width modulated (PWM), pump charging, etc. PWM types of controllers operate at a very high switching frequency and they can cause interference with certain types of loads, such as radio equipment. These types of controllers do, however, charge the battery more efficiently.

'Float' charging is the use of a reduced end-of-charge set point in order to maintain the battery near a full state of charge. Typically the controller holds the voltage of the battery constant at the float set point by adjusting the current. In general, float charging is used to compensate for self-discharge when there is no load; in PV systems, it may also be used for very shallow cycling, such as that which occurs during summer with a system sized for winter operation. Float set points are not high enough to cause the gassing necessary for mixing the electrolyte or ensuring full charge.

Voltage charge set point

The voltage set point of a charge controller should not be too high or there will be excessive gassing. Conversely, it should not be too low or the battery will never be fully charged which, for lead–acid batteries, will result in sulphation, loss of capacity and shortened life. Most controllers have variable threshold points which can be selected by a continuous variable resistor or by dip switches. These should be adjusted carefully.

Exact battery control points are specific to battery types and may even vary from manufacturer to manufacturer. The charge set point is also influenced by other factors – the most important being the controller type, battery temperature and rate of charge. While we cannot give a universal set of rules for setting the voltage at which the controller should regulate the current to the battery, Table 7.1 gives a first indication for different battery types and for the two main controller types. Variations for different temperature and charging systems are discussed in following sections.

Compensation for rate of charge

As we have seen in Chapter 6, the relation between the voltage of a battery and its state of charge varies with its rate of charge. The maximum voltage set point for both lead–acid and Ni–Cd batteries will thus be influenced by its rate of charge. Figure 7.2 illustrates the end-of-charge voltages for different charge rates for a typical lead–acid tubular battery.

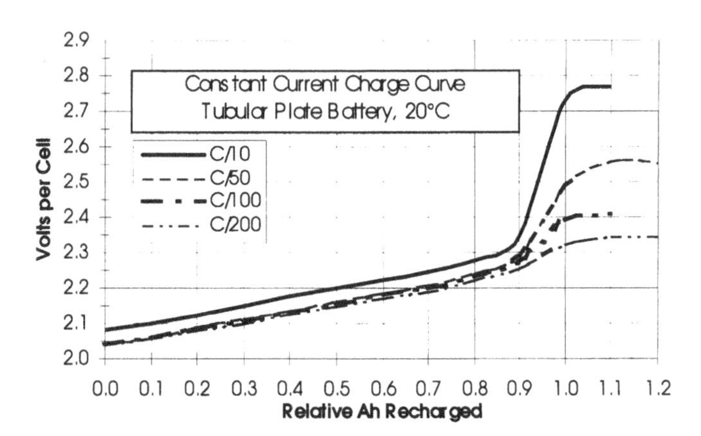

Figure 7.2. End-of-charge voltages of a lead–acid battery for different charge rates[2]

For low-current charge rates, which are typical of PV applications, the proper set point should be set lower than the one specified by the manufacturer (usually given at 10 hours charge) so as not to overcharge the battery. For the tubular lead–acid battery shown in Figure 7.2, the charge set point of a one-step controller (set halfway up the steep portion of the curve between 90 and 100% charge returned) and for a PWM controller (set a quarter of the way up this end-of-charge curve) are as shown in Table 7.2.

Table 7.2. Charge set points

Charge rate	One-step controller	PWM controller
C/10	2.50 V/cell	2.40 V/cell
C/50	2.40 V/cell	2.33 V/cell
C/100	2.35 V/cell	2.30 V/cell
C/200	2.30 V/cell	2.27 V/cell

Compensation for temperature

The charging characteristics of lead–acid batteries change with their temperature. The colder the battery, the lower the rate of charge it will permit. For example, a particular lead–acid tubular battery that is half charged may only accept a

Table 7.1. Nominal charge set points for 12 V batteries: ref. adapted from [1] and [2]

Battery type	Typical set points (V) for various batteries and controllers at 25°C and *C*/20 charge: Float voltage	One-step controller	Switching controller	Equalization
Flooded lead–acid low-antimony, flat plate	13.8	14.3–14.8	14.1–14.6	15.3 V for 3 h
Flooded lead–acid tubular plate	13.8	14.4–15.0	14.1–14.4	15.3 V for 3 h
Low-maintenance lead–calcium	13.5	14.1–14.3	14.1–14.3	Not recommended
Lead–acid gel, sealed	13.2	14.1	14.1	Not recommended
Lead–acid AGM, sealed	13.2	14.1	14.1	Not recommended
Nickel–cadmium, sealed	Constant current at C/10	Constant current only	Constant current only	Not required

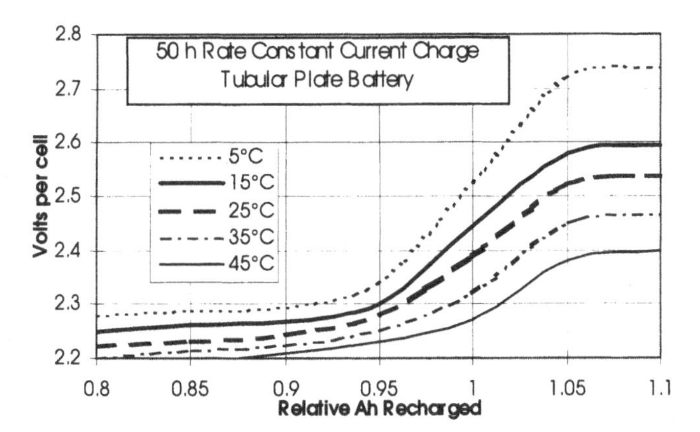

Figure 7.3. End-of-charge voltages of a lead–acid battery for different temperatures[2]

charge of 2 A at –20°C, while it will accept 25 A at +25°C. At a constant charge rate of C/20, the same battery will need to attain 2.8 V at –20°C to be completely recharged, as opposed to 2.5 V at +25°C (see Figure 7.3). Moreover, as we have seen earlier, at lower temperatures, lead–acid batteries are also prone to freezing.

Most charge controllers equipped with temperature compensation will adjust the charge set point to allow proper charging of the battery. To prevent excessive gassing, this set point will be reduced at higher battery temperatures. Conversely, it will be increased at lower battery temperatures so that the battery may be sufficiently recharged.

Equalizing the battery

As the cells in a battery will probably age at different rates, normal charging may lead them to develop different SOCs at the end of a recharge. From time to time it is good practice to give a much larger amount of overcharge than normal, in order to let the weaker cells charge up fully. This is called an *equalizing charge.*

Applying an equalizing charge is particularly important after the system has been allowed to be deeply discharged. If a lead–acid battery is deeply discharged, it will need to reach a somewhat higher end of charge voltage than normal in order to be fully recharged.

Lead–acid batteries subjected to float charging also tend to stratify and need to be subjected to periodic equalization charges. This allow gas bubbles to stir up the sulphuric acid, which otherwise tends to concentrate at the bottom of the cell (*stratification*). This is particularly important in high-latitude, cold-climate applications where the battery may be left on float charge during winter. If the battery stratifies, the top layer may freeze.

Equalization can be done easily by allowing the controller to pass its normal charge set point for a limited time, usually between three to five hours, as indicated in Table 7.1. However, equalizing of the battery will produce gassing and thus a loss of water in the electrolyte. It should be performed only when it is possible to replenish the water loss with distilled water.

In a gel or absorbent glass mat (AGM) battery, the equalization charge must not exceed the maximum rate at which the battery can recombine the oxygen and hydrogen gas produced during overcharge. Since the maximum recombination rate is mainly influenced by the temperature of the battery (slower at low temperature), equalization of these batteries at cold temperatures is usually not recommended. If the charge current is too high, the internal pressure will build up and acid spray will be vented out, resulting in loss of capacity and irreversible acid loss.

Some controllers are designed to automatically adjust for periodic equalization charges. They may also give an equalization charge when they detect that a very deep discharge has occurred. These controllers are often used in PV/genset hybrid systems since it is easy to allow the genset to run for a little longer past its normal charge set point.

With an autonomous PV–battery system the equalization process may be impossible if the PV array is too small to overcharge the battery adequately. Usually this occurs during winter; the controller should ensure that the battery is equalized during the spring when sufficient charge current is available. Another approach is to visit the site with a transportable power source in order to perform equalization.

ENERGY MANAGEMENT STRATEGIES FOR AUTONOMOUS SYSTEMS

A proper energy management strategy for a PV system will ensure that the energy produced by the PV array is used at its maximum whenever the sun is shining, while still making sure that the system is able to power the load efficiently when the sun is not shining. In cold-climate operation this means that the system must be able to provide power even when sunlight availability is scarce and when cold temperatures adversely affect the efficiency of the components used in the system. In the next sections we will give guidelines for proper energy management strategies using different PV power systems.

Strategies for autonomous PV–battery systems

Autonomous PV–battery systems are usually used when the demand for electricity is small, typically less than a few kWh/day and when the period of use corresponds to the period of greatest available sunlight. They are particularly well suited for summer vacation residences, weekend cottages and cabins. This means that PV systems may be needed to supply power for brief periods and then left in a charging mode for long periods. It also means that they may be left unattended for very long periods during the winter months.

Discharge strategy
In a standalone PV–battery system, discharge control usually operates in two steps:

- a low-voltage alarm which may be coupled with the shedding of unimportant loads;
- a complete disconnect of the loads from the power source.

In warm climates, the load disconnect point is usually set at 20% SOC when deep-cycling batteries are used. In cold climates, where the battery is prone to freezing, this level must be set higher and will depend on the battery type and the coldest temperature that the battery will encounter during the year. As an example, a typical deep-cycle lead–acid battery will freeze at 50% SOC when the temperature is –20°C.

The low-voltage alarm and low-priority load disconnect is usually set at a level significantly higher than the complete load disconnect. By disconnecting low-priority loads, the remaining loads will be powered for a longer time before the system needs to be shut off. There is an added benefit because the battery has greater capacity at lower discharge rates; reducing the load by 50%, for example, will more than double the time remaining before disconnect.

Charging strategy
The main strategy for charging a battery in a stand-alone PV–battery system is to recharge it as fast as possible without overcharging it and causing excessive gassing. Since the PV array has a variable output, constant current control of the battery charging is usually not possible. Instead, the full array current is used to charge the battery until a certain voltage is reached, at which point the array current is cut off completely, reduced by switching off part of the array or pulsed in an attempt to keep the battery voltage as constant as possible. As we have seen, this cut-off voltage depends on the type of controller used and the maximum rate of charge the battery will experience.

When the battery is oversized in order to supply the load for some time during low-insolation periods, the charge current of the battery will be relatively low. Typically, the battery may take from 50 to 200+ hours to be recharged completely. This is particularly true in high-latitude applications where the battery is often asked to provide the load without any input from the PV array for a few weeks to a few months.

For low-current charge rates, the proper set point is lower than the one specified by the manufacturer (usually given at 10 hours charge) – see the section on optimizing battery operation. However, when the PV array is oversized to meet the load during the worst-case insolation period, the maximum charge current (during sunny periods) will be much higher than the average charge current. If the charge set point is set on the basis of this maximum current, the battery will be in danger of overcharge whenever the array current is lower. On the other hand, if the charge set point is selected on the basis of a low current, much of the array's output will be wasted. A compromise must be found between these two, typically in the neighbourhood of the set point for charge rates of C/20 to C/50.

In areas north of the Arctic Circle, when equipment needs to be powered year-round, the PV system may be designed so that the battery bank will supply the charge without being recharged for many months while keeping its state of charge above the minimum allowed. The PV array will then be small relative to the battery bank, but sufficiently large that it slowly brings the battery to the full state of charge over the summer months, ready for the next dark period. In this type of system, the charge controller may have to be designed with special features to have the PV array address individual battery cells or groups of cells in succession to ensure that each is fully charged.

Strategies for PV–genset hybrid systems

While the operation of PV–genset hybrid systems is discussed in Chapter 9, there are several points specifically related to the charge and discharge of the batteries in hybrid systems that merit attention here.

Discharge strategy
The discharge algorithm in a hybrid system normally has two primary discharge set-points in order to control the genset start-up and manage the load. Both of these functions are usually based on battery SOC.

The first 'set point' specifies the maximum depth of discharge: at this point, the genset is started in order to recharge the battery. This is based on the battery type and is a trade-off between cycle life and cycle depth. In a simple battery/genset system (without PV), the ideal trade-off can be identified, with a single value emerging as to the optimum discharge level. In a hybrid system, however, there are intermediate cycles due to recharging by the PV array during sunlight hours. In cold climates this is further compounded by the following factors:

- At low SOC, the battery is more prone to freezing.
- The diesel genset needs to start-up every once in a while.
- The genset thermal output may add to the battery capacity.

A study carried out by Integrated Power Corporation and Westinghouse Electric suggested that a level between 25% and 50% SOC is the optimum point when the generator must be started.[3] Depending on exact system parameters, this level should permit limited start-up from the generator as well as adequate runtime while permitting maximum use from the PV array. This optimal SOC level will be greatly influenced by the temperature to which the battery will be subjected, however. As with the autonomous system, the SOC should never reach a level where the battery can freeze. This can be avoided by starting the generator when the battery reaches a certain SOC level calculated for the worst temperature. However, we have seen that this SOC may need to be set at a fairly high level if the batteries are left unprotected.

The second set point is a low-voltage disconnect, set at a voltage or state of charge lower than the genset start set point. This ensures that if the genset fails to start for some reason, the battery will not be over discharged to the point where it would be damaged by cell reversal or freezing. Sometimes a third set point, set at a voltage between the genset start set point and the low-voltage disconnect, is used to trigger alarms and shed non-essential loads before complete system shut-down.

Charge strategy

The primary strategy when using a genset to recharge a battery is to charge it in the shortest possible time (in order to keep the genset run time to a minimum) without endangering the integrity of the battery.

In a simple cyclic battery–diesel system, the genset is designed to recharge the battery at a constant current until the voltage reaches a certain voltage, then switch to a constant-voltage mode and taper the current appropriately for a fixed time. This constant voltage set point is typically set at a level low enough not to provoke gassing of the battery. This strategy is not very efficient since during constant voltage charging the battery accepts little current and the genset is used at a low capacity level. Thus, in a typical PV–genset hybrid system, the genset is allowed to recharge the battery up to a certain cut-off voltage which usually represents 90% SOC (a little lower than the constant voltage set point) and the generator is then shut-off. The PV array then finishes the charging of the battery.

Because the genset is usually sized to minimize its 'on' time, the charge current will be high, in the region of C/5 to C/10. Thus, the voltage at which the genset will be turned off will also be relatively high. For example, for the tubular lead–acid battery in Figure 7.2, the cut-off voltage for the diesel generator should be set at 2.3 V in order to ensure that the battery is 90% charged. Depending on the relative size of the PV generator *vis-à-vis* the battery bank, the set point at which the PV array takes over will be chosen according to the maximum charge current it can deliver and the controller type (see above).

USING PV ON A REMOTE GRID

On a remote grid, a PV system can be connected to the grid near the point of demand (individual decentralized grid-connected PV systems) or at the point of production (PV central station). Its use will reduce consumption of fuel and, to a lesser extent, reduce generator runtime. We have seen in Chapter 4 that the economics of installing PV on a diesel grid depends on various parameters, which may or may not favour its installation. Its operation, on the other hand, is relatively straightforward. Because the energy produced by the PV array is variable and not on demand, it should be used whenever it is available and not stored in batteries. The PV generator then acts as a negative load, decreasing the overall load as it is feeding energy into the grid. The grid acts as a very large battery, absorbing all of the production of the PV array.

When more than one diesel generator is on line, the contribution from the PV system may also permit the shut-down of one of the generators, thus reducing runtime and maintenance costs. However, this may increase start/stop frequency, which can have adverse effects on generator lifetime.

An analysis conducted by CANMET's Energy Diversification Research Laboratory in Canada showed that as a result of closely matched insolation and diesel plant load during the summer period, PV can be implemented to a ratio of 50% of the peak load in Canadian remote communities, without dumping substantial energy.[4] In this case, depending on the price of fuel and other economic factors, the avoided cost benefit ranged from 1.30 to 3.30 USD/W_p.

The most critical energy management issues when using PV on a remote grid are that the power quality of the inverter must match the power quality of the gensets and that the PV system must shut down when the genset is shut down in order not to back-feed electricity onto a line that is supposedly non-powered. These issues are addressed by using special grid-tied PV inverters.

REFERENCES

1. Photovoltaic Design Assistance Center (1994). *Battery and Charge Controller Workshop for Photovoltaic Systems.* New Mexico State University, NM: Sandia National Laboratories.
2. Spiers D, Royer J (1998). *Guidelines for the Use of Batteries in Photovoltaic Systems.* Helsinki: Neste Advanced Power Systems; Varennes, Quebec: Natural Resources Canada.
3. Danley DR, Integrated Power Corporation and Westinghouse Electric Corporation (1990). *Development of Photovoltaic System Design Incorporating Advanced Control Algorithm.* Ottawa, Ontario, Canada: Efficiency and Alternative Energy Technology Branch, Energy, Mines and Resources Canada
4. Usher E, Martel S (1995). *Avoided Cost Benefits of Photovoltaics on Diesel-Electric Grids.* Varennes, Quebec, Canada: CANMET Energy Diversification Research Laboratory

III Designing photovoltaic systems for cold climate regions

8 Stand-alone PV-only systems

Alf Bjørseth[1], *Knut Hofstad*[2] *and Fritjof Salvesen*[3]
(ScanWafer AS[1], Norwegian Water Resources[2] and Energy Administration, KanEnergi AS[3])

One of the most common and cost-effective applications of photovoltaics is the stand-alone system, which typically employs a photovoltaic array to charge a battery. The battery furnishes power at night and when skies are cloudy; the system may include a charge controller to regulate operation of the battery and an inverter to convert the DC electricity of the battery into AC electricity, for those applications requiring it. In some systems, such as those providing power for water pumping, the battery is not necessary and the PV array may be connected to the load directly or via an electronic controller.

In many cold-climate regions, PV–battery stand-alone systems are attracting increasing attention. This is due to the apparent advantages of such systems, even with the limited winter solar resource at high latitudes:

- They provide independence from grid connection and are particularly suited to applications where grid connection would be costly.
- They are a highly reliable technology, even in very harsh environments.
- They operate automatically and do not need any human intervention under normal operation.
- They are easy to monitor and operational disturbances can be detected at an early stage.

In cold-climate countries such as Norway, typical applications of stand-alone PV systems include:

- coastal lighthouses;
- remote cottages and mountain huts (Norway has 60,000 such PV installations);
- remote residences;
- telecommunications systems;
- remote monitoring systems.

These installations have led to the development of techniques and tools for the design of stand-alone PV systems. The standard procedure is conceptually straightforward, but designing the ideal system for a given load and site is not trivial. Cost and reliability must be balanced. Properly used, computer simulation and modelling packages can identify the least-cost configuration that achieves a desired level of reliability.

DESIGN OF STAND-ALONE PV SYSTEMS FOR COLD CLIMATES

Is a stand-alone system appropriate?

Before designing a stand-alone photovoltaic system, ensure that it makes sense. As mentioned in Chapter 4, photovoltaic stand-alone systems tend to be ideal for off-grid applications requiring a small, reliable, low-maintenance power supply. If the application under consideration does not fit this description, revisit Chapter 4 and examine alternatives: for example, grid connection, a genset or a PV–hybrid system may be more appropriate if a large amount of power is required.

Design method

Designing a photovoltaic system for a cold climate is not unlike designing a photovoltaic system for a warm climate, though the differences in climate lead to substantially different designs. Ideally, sizing and simulation software will be used to search for the optimal design, but a manual approach can yield acceptable results and serve as a verification of the software. Design proceeds in six steps:[1]

1. *Estimate the available solar energy*. For stand-alone systems, the 'worst-case' availability of solar energy is critical. Sunshine is usually scarce during winter but abundant during the rest of the year. Winter months are crucial for year-round operation; the last months of summer are most important for summer-only systems.

   As mentioned in Chapter 3, solar energy data is generally available in the form of monthly averages for sunshine on a horizontal surface. This data will have to be adjusted to account for the tilt angle and orientation of the array. Furthermore, these average values mask differences between good and bad years. During bad years, the total insolation during the month may be far lower than would be expected from an average calculated for that month over, say, 30 years. Thus, it is prudent to consider not just average solar radiation for the least sunny month, but also *historical lows*. The resulting estimate may be further derated to account for those periods when the array will be shaded by snow or ice (see Chapter 5), mountains, trees or buildings (see Chapter 3).

2. *Determine the average total load.* List all the DC and AC loads. For each load, multiply the power requirement for the load by the percentage of the time that it will be used (i.e. its 'duty cycle'). For AC loads, divide the product by the efficiency of the inverter. Sum the results: this is the average total load.

   At this point, search for ways to reduce energy consumption, through more efficient electrical loads (e.g. fluorescent instead of incandescent lamps) or by switching appropriate appliances to other fuels (e.g. propane instead of electricity for refrigeration). Investment in energy-efficient devices will often be recouped through reductions in the size and cost of the photovoltaic system.

3. *Size the battery.* Determine an appropriate level of battery 'autonomy', or period of time that the battery could power the average load, were the battery fully charged and then the PV array disconnected. The autonomy must reflect a number of factors, discussed below. The size of the battery, in watt-hours, is the product of the average load, in watts, and the number of hours of autonomy. Similar considerations apply to pumping systems incorporating water storage.

4. *Size the array.* The array must be sized on the basis of the power it provides during the least sunny period (i.e. month) of system operation. This should exceed, by a margin of 10% to 40%, the average total load plus any inefficiencies in the controller, battery and wiring. (Note that most simple charge controllers will not operate the array at the voltage at which the array generates the most power, leading to further losses that must be accounted for.) The margin by which the array power should exceed the load depends on:

   - the desired reliability of the system (higher reliability demands larger margins);
   - whether the estimate of solar radiation is an average (a large margin is used) or a historical low (a small margin is used);
   - the size of the battery (larger batteries permit smaller margins).

   Note that at high latitudes the use of seasonal storage permits *negative* margins, greatly reducing array size and cost.

5. *Evaluate hybrid system options.* In some cases, a PV–hybrid system may make more sense than a PV-only system. Using the PV-only system for comparison, consult the guidelines in Chapter 9.

6. *Choose components and configuration.* Select the modules, batteries, electronics, etc. with due consideration for the cold-climate issues discussed in Chapters 5, 6 and 7.

RELIABILITY OF STAND-ALONE PV SYSTEMS

Types of failures

When reliability is discussed, it is important to differentiate between different types of system failure. Consider three hypothetical mountain-top PV systems for telecommunications antennas. All three have failed: they can no longer provide power to their antenna. The first has a broken charge controller. The second was designed for typical weather conditions; unfortunately, recently the weather has been especially overcast. The third has experienced heavy rime icing and the array is encased in ice. In all three cases, the array provides insufficient power to charge the battery. Yet they have failed in three different ways: catastrophic failure, exceptionally dark weather and freak occurrence of icing.

Catastrophic failures, which occur when a component malfunctions, a wire becomes loose or a fuse blows, are very rare in PV systems. When people say that PV is reliable, they are referring to the low incidence of these catastrophic failures.

Failures due to exceptionally dark weather occur when a system has been designed assuming a certain level of sunshine, and then the real level of sunshine falls below this level. These failures can be minimized, but never eliminated. For example, a PV designer may examine weather data for the past 30 years and determine that in 97% of those years, the amount of sunshine during the darkest month did not fall below a certain level. The designer then sizes the system such that it will function at this level of sunshine. The very next year, December is exceptionally dark – it is one of those 3% of years when the sunshine falls below the design level – and the system fails. When some people say that PV is not reliable, they are referring to the impossibility of eliminating periods of very dark weather that will cause system failure.

Freak occurrences of icing are a concern only at certain sites in mountainous or polar regions (see Chapter 5). Rime ice may block sunlight to the array for periods of days, weeks or even months, making it an important design consideration at these sites.

Measuring reliability: loss-of-load probability

Sometimes it is useful to measure reliability. The user can then compare the reliability of different systems and evaluate whether a system is sufficiently reliable for his or her purposes. The 'loss-of-load probability' is a common yardstick for stand-alone PV systems. It is defined as the fraction of the time that a PV system is unable to provide power to the load. For example, if a PV-powered light is supposed to operate for ten hours each night, but on average the PV system can furnish power for only nine hours, then the loss of load probability would be 10%. Reliability is inversely related to the loss-of-load probability; a highly reliable system has a very low loss-of-load probability.

If a system has already been in operation for a long time, its loss-of-load probability can be easily calculated from monitored data. Predicting the loss-of-load probability of a hypothetical system is more difficult, but also more useful. There are two approaches: simulation and analytical methods.

Simulation is generally the most accurate way to calculate the loss-of-load probability. A computer is used to predict the operation of the proposed photovoltaic system over many years; the number of hours when the system failed divided by the total hours of demand yields the loss-of-load probability. This requires immense quantities of hourly or daily weather data, which may be unavailable to the designer. Data synthesized by a computer can be substituted for monitored data.[2]

Analytical (or 'stochastic') methods eliminate the need for large quantities of weather data. Using a little mathematics and a few weather statistics, they estimate the probability that a protracted period of bad weather will cause system failure. These methods can even determine the optimal battery size for a given array. Unfortunately, they are not as accurate as simulation.[3]

Both simulation and analytical methods ignore catastrophic failure and reductions in array output due to snow and ice accumulation; only the variation in the weather is considered. Since catastrophic failures are rare and will affect different systems to roughly the same extent, they can be neglected without penalty. If snow accumulation is a concern, solar energy input should be derated. For sites prone to serious rime icing, calculating the loss-of-load probability requires estimating the probability of rime – which is generally impossible.

Reliability and system cost

If a more reliable photovoltaic system is desired, the array, the battery or both the array and the battery must be enlarged. Not surprisingly, this raises system costs (see Chapter 4). Note that the law of diminishing returns applies here: the more reliable a system is to begin with, the more costly it is to make it incrementally more reliable. That is, reducing the loss-of-load probability from 2% to 1% will cost much more than reducing it from 3% to 2%.

The designer of a stand-alone PV system should determine the level of reliability required by the user and then size the system accordingly.[4] The cost of a power failure must be weighed against the cost of improving reliability. Undersizing the system will lead to an unacceptable number of system failures, while oversizing the system will lead to inflated system costs. Cottages and other domestic systems may permit a high loss-of-load probability – perhaps in the neighbourhood of 5%. At the other end of the scale, telecommunications systems may require a loss-of-load probability smaller than 0.1%.

BATTERY AUTONOMY FOR COLD CLIMATES

How much battery autonomy should a stand-alone system have? This depends on five factors:

- *The reliability required.* Higher levels of autonomy cost more, but can improve reliability; the level of autonomy should reflect an appropriate trade-off between cost and reliability, as discussed above.
- *Weather variability.* The array is sized to produce enough power for average levels of sunlight during the worst-case period. During this period, however, the sun will not always shine with the same strength: at night it will not shine at all; some days it will be bright; some days it will be cloudy. The battery must buffer these variations and, in climates where this variation is exaggerated, a larger battery will be required. Many mountainous regions are known for persistent cloudy weather.
- *Battery temperature.* Battery autonomy is commonly quoted with no consideration for battery temperature. This is extremely misleading in cold climates. As discussed in Chapter 6, a cold battery cannot be discharged as deeply as a warm one. Forty days of lead–acid battery autonomy is really only about 25 days of battery autonomy at –20°C. This can be conservatively accounted for by choosing the battery autonomy on the basis of a battery at 25° C, then enlarging the battery so that it provides this level of autonomy at the predicted minimum battery temperature.
- *Severe rime icing.* At some polar and mountainous sites, rime ice can encase the array for weeks or months; during this period, the battery will have to supply almost all the

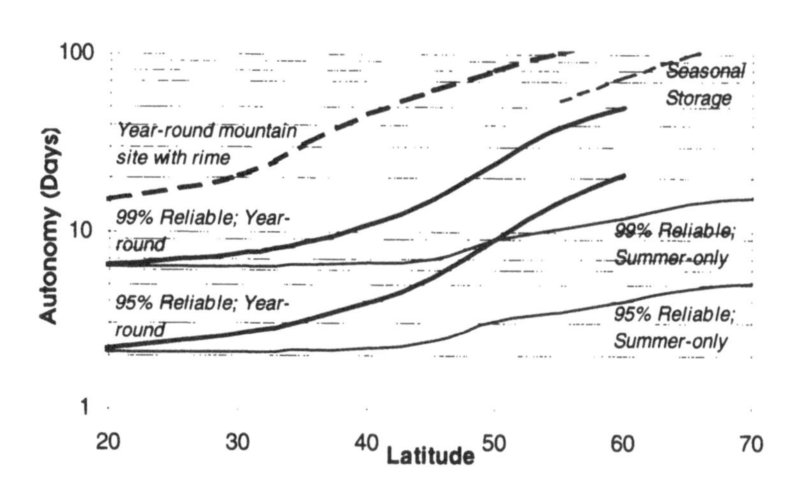

Figure 8.1. Typical levels of autonomy for PV-only systems with lead-acid batteries

power. The battery autonomy must reflect the expected duration of rime icing.

- *Seasonal storage.* At high latitudes, the solar radiation available during the worst-case month may be so minimal that an array that could produce sufficient power to meet the load during this month would be prohibitively expensive. In such cases, it may be cheaper to oversize the battery, such that it can power the load for these least sunny months, and keep a moderately sized array. For example, lighthouses in southern Norway are powered by stand-alone PV systems with 60 days of battery autonomy; lighthouses in northern Norway (up to 71°N) have 120 days. Note that when seasonal storage is used, lower panel tilt angles may be optimal.

Typical levels of autonomy for different systems and sites are shown in Figure 8.1. Lead–acid batteries in unheated enclosures with little insulation are assumed; since cold temperatures affect them less, Ni-Cad battery banks are usually smaller than indicated here, especially at high latitudes. Summer-only systems (i.e. April through September in the northern hemisphere) are compared with year-round systems. The year-round curves assume no seasonal storage and are not shown for latitudes above 60°; at these latitudes array sizes become so large that seasonal storage is usually employed. The curve for seasonal storage is very approximate, since the level of seasonal storage will depend greatly on the array sizing. The curve for mountain sites with rime assumes lower temperatures than the other year-round curves; increasingly severe rime accumulations are assumed at higher latitudes. In contrast to these PV-only systems, PV–hybrid systems usually have one to four days of autonomy.

OPTIMIZING THE SIZING OF STAND-ALONE SYSTEMS

Stand-alone systems are quite simple. Many people therefore assume that their design and sizing is simple. In part, this is true: it is not hard to find a design that works. In contrast, it is difficult to identify the optimal design, that is, the design which achieves the required system reliability at the least cost. Complications arise due to the natural variation in sunshine over a period of time; predicting a system's performance over several weeks of changing weather is not intuitive. Proper design tools, modelling and simulation are required to ensure optimal design. As discussed in Chapter 11, such tools are widely available and their use is strongly recommended.

The same principle underlies most of these tools and techniques.[5] Assume that the designer has been given the task of sizing a stand-alone PV system for a certain load. The PV system will have to function at a certain level of reliability. There are, in fact, an infinite number of different combinations of array and battery size that can achieve this level of reliability. For example, a big array and a small battery can have the same reliability as a small array and a big battery.

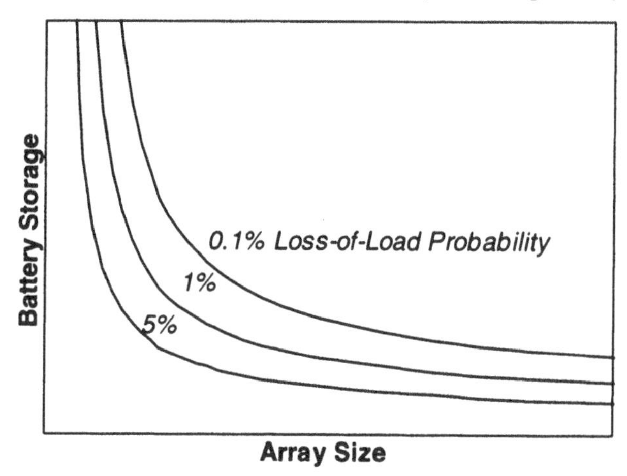

Figure 8.2. Combinations of array and battery having the same reliability

These combinations of array and battery describe a curve on a graph of array size versus battery storage. All combinations on this curve have the same loss-of-load probability. If this is done for different values of loss-of-load probability, a family of curves results (see Figure 8.2).

The first concern in optimization is that the system be located somewhere on the appropriate reliability curve. For example, assume that a system with a loss-of-load probability of 0.1% is desired. If the design is not on the 0.1% loss-of-load probability curve, it is either undersized, and will fail too frequently, or is oversized, and costs too much. Neither is optimal.

Just because a system is on the right reliability curve does not make it optimal. Only one point on this curve is optimal: it is the combination of battery and array which minimizes cost. Identifying this point is the second concern of optimization. Proceed by overlaying lines of constant cost on the reliability curves; the lines of constant cost represent all the combinations of array and battery that sum to the same cost. Most constant-cost curves will either not intersect the desired reliability curve or intersect it in two places. One constant-cost curve will, however, intersect the desired reliability curve at only point; this point identifies the optimal system (see Figure 8.3).

Locating the constant reliability curves is not straightforward. Some formal method and sophisticated tool is required; *ad hoc* sizing methods cannot be expected to find the optimal system.

In cold climates, optimization is complicated by low temperatures and, at those sites susceptible to it, rime and snow accumulation. Few analytical optimization methods have been adapted to account for battery temperature, which will vary in typical stand-alone systems. Simulation can account for battery temperature directly, but many simulation packages do not have models for battery temperature, as discussed in Chapter 11. Furthermore, the above analysis is predicated on the dominant cause of system failure being a prolonged spell of cloudy weather; when rime is the dominant cause of failure, the approach breaks down. The approach is still useful, however, since it establishes a lower bound on array and battery size.

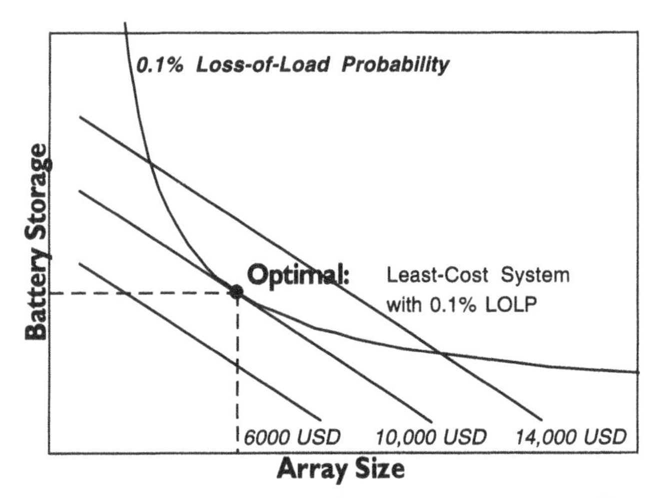

Figure 8.3. The optimal system is the least-cost system with a given reliability (currency in USD)

BARRIERS TO GREATER USE OF STAND-ALONE PV SYSTEMS

Stand-alone PV systems are simple, reliable, versatile and environmentally friendly, yet there are many applications where they are not presently used. Why is this?

The most important reason is cost. PVs cannot compete with the grid and for large power requirements other technologies are cheaper. Yet even when PVs are economically viable, they are sometimes not used, suggesting that there are other barriers.

Many people are not familiar with PV technology and are unwilling to try it. Some have also had unfavourable experiences with other types of stand-alone power systems, such as gensets or wind turbines. Others are concerned about maintenance costs: the lifetime of a battery can be very short if used improperly. The lack of experience increases the uncertainty. People resist changes in habit. Even though a proposed system may not imply any reduction of energy quality or comfort, people may resist it.

Often utilities and municipalities are even more intransigent than individuals. Accustomed to traditional means of electricity generation and distribution, they are reluctant to consider alternatives. Existing regulations may not accommodate stand-alone PV systems.

To overcome these barriers, people must be provided with realistic information about experiences and solutions. Monitored demonstration systems that can be visited and inspected by the public are useful for this purpose.

REFERENCES

1. Natural Resources Canada (1991). *Photovoltaic Systems Design Manual.* Ottawa: Natural Resources Canada.
2. Klein SA, Beckman WA (1987). Loss-of-load probabilities for stand-alone photovoltaic systems. *Solar Energy*, 39(6) pp.499–512.
3. Egido M, Lorenzo E (1992). The sizing of stand alone PV-systems: A review and a proposed new method. *Solar Energy Materials and Solar Cells*, 26, pp.51–69.
4. Sandia National Laboratories (1995). *Stand-alone Photovoltaic Systems: A Handbook of Recommended Design Practices.* Albuquerque, NM: Sandia National Laboratories.
5. Gordon JM (1987). Optimal sizing of stand-alone photovoltaic solar power systems. *Solar Cells*, 20, pp.295–313.

9 PV–hybrid systems

Jito Coleman
(Northern Power Systems)

Hybrid systems, as mentioned in Chapter 2, are power systems containing more than one type of electric generator. For example, a PV–hybrid system may contain a diesel genset and/or a wind turbine in addition to a PV array. PV–hybrid systems are well suited to cold climates because they address the difficulties of having abundant sunshine in the summer but little in the winter: whenever sunshine is not available, the other electric generator provides power. Compared to PV-only systems, PV–hybrid systems are more complex, offer more flexibility and demand more decisions related to design and control. Often they are used for applications requiring more power than that supplied by typical PV-only systems.

Hybrid systems appear in many different forms, from large, complex systems for remote communities to small PV arrays coupled to small wind turbines. In this chapter, we focus on the cold-climate issues related to PV–hybrid systems for off-grid loads, such as remote telecommunications systems and remote residences.

TYPES OF PV–HYBRID SYSTEM

PV-only systems must be sized to provide power to the load at that time of the year when sunshine is least available. Arrays and batteries must be large, so that what little sunshine is available during winter can be captured and stored. If some other electric generator can be used during the winter, however, the array and battery size can be reduced, lowering costs. Moreover, system reliability is improved, since the probability of both the PV array and the second generator simultaneously failing to power the load is lower than the probability of just one of the two failing. Thus combining different types of generators improves reliability while decreasing cost.

PV–hybrid systems can be categorized by the nature of the generators they contain:

- Systems containing only site-dependent, time-dependent renewable-energy generators.
- Systems containing one or more on-demand fossil-fuel-powered generators.

Typical renewable-only hybrid systems include a battery, a PV array and a wind turbine. In many areas, winds tend to be stronger in the winter than the summer, so that the PV array and the turbine complement each other. Unfortunately, just as one cannot predict the minimum amount of sunshine available in, say, December, there is no way to predict how little wind might blow. There is always a small chance that a renewable-only hybrid system will fail because of less sun and wind than anticipated during design; furthermore, systems sized for those times when wind and sun are scarce will be oversized under normal conditions.

In contrast, the output of a fossil-fuel generator can be controlled by turning it on or off. As a result, the PV array need not be sized on the basis of the sunshine available during winter, since the fossil-fuel-powered generator can be turned on whenever needed. As long as the additional generator does not break or run out of fuel, the system will be perfectly reliable. This is the advantage of *power on demand.*

The most common PV–hybrid system contains a battery, a PV array and a genset. The system makes sense financially: PV arrays are costly to purchase but cost little to operate and maintain, while gensets are the reverse. When combined, the genset reduces the size – and therefore the cost – of the array, while the array reduces fuel use and thereby lowers operating costs.

PV–HYBRID SYSTEM COMPONENTS

Renewable-energy generators

Wind systems

Wind systems can be a good choice if the site has a strong wind resource. Wind resources tend to be characterized on the macro scale from data that has been compiled by various national or provincial organizations. Local or micro-scale impacts must also be assessed to ensure good turbine siting. *In situ* measurements, while not absolutely necessary, certainly enhance the level of knowledge of the wind resource at the specific site. Siting turbines in mountainous areas is much more demanding than siting in flatter areas, where there are fewer topographical features to affect wind patterns.

Low temperatures can be a significant concern for wind turbines, as the metals used in these devices will become brittle at low temperatures and be subject to impact failure and accelerated fatigue. Selection of appropriately rated low-temperature turbines is necessary when extreme or prolonged low temperatures are expected. The status of the wind turbine industry has matured to the level that good-quality equipment is available for cold-climate applications.

Concern about icing of the wind turbine in cold climates is justified. Wind-turbine blades can become coated with ice, affecting their aerodynamics to the point that they will not turn. Sites with severe or prolonged icing may be poor candidates for wind systems. In general, the regions of the world that should be avoided for icing are coastal sites, where the abundance of water vapour can create severe icing. At these sites, the system will be prone to icing for extended periods of time, preventing the turbine from being able to supply energy for many months. At mountainous sites, the conditions that cause icing can last from several hours to a few weeks. It has been reported that on sites particularly prone to rime icing, the annual energy output of a wind turbine can be decreased by up to 17%. At many other sites, icing is a short-term phenomenon that will not greatly affect the output of the wind turbine.

Icing so severe that it poses a structural hazard is rare, but, where it might exist, an appropriately designed turbine and tower must be chosen in order to avoid damage to the system.

De-icing provisions for wind turbines have not been in use and have not been justified in the environments where wind turbines are currently being employed.

Hydro power

Hydro power is occasionally considered as a generator in PV–hybrid systems. Micro-hydro systems are appropriate when sufficient hydro resources are very near the site. Hydro opportunities are very site dependent, typically requiring the facility to be located within 500 m of the site. Evaluation of the hydro resource should include monthly evaluation of the water flow potential to verify that the water resource complements the solar resource. If hydro power is available on a year-round basis, there may be no reason to use solar power at all. While technically feasible, there are very few hybrid solar/hydro facilities, and the ones that do exist have a specific set of energy and operational requirements that justify their use.

Hydro facilities may have problems in cold environments if the waterways, raceways or control gates become iced up. In addition, flow rates may be much lower in the winter months, precisely when the hydro facility is expected to power the load.

Fossil fuel-powered generators

Fossil fuel-powered generators are usually the first option considered for PV–hybrid systems. This is because traditionally the fossil fuel-powered generator was the primary source of energy for remote applications. When this option is considered, there are two major issues: the type of fuel and the conversion method. In cold climates and remote locations both these issues are important.

Thermal electric generators

Thermal electric generators (TEGs) are energy-conversion devices that burn gaseous fuels, such as propane and natural gas, in a combustion chamber that provides heat to one side of a thermocouple junction. The thermocouple junction produces an electrical potential based upon the difference in temperature between the hot and the cold junctions. TEGs come in unit sizes from 10 to 120 W, which can be paralleled to provide larger power systems.

TEGs are very reliable as there are no moving parts, but they are only 3–4% efficient in converting fuel to electricity. TEG output voltages are matched to typical telecommunications equipment requirements (12, 24 or 48 V DC). The most unreliable aspect of the TEG is ignition, which is generally accomplished by piezoelectric igniters on each TEG cell. Turning TEGs on and off is not recommended and typically TEGs are left running once started. For this reason, in PV–TEG hybrid systems the TEG is usually started at that point in the autumn when the solar resource is inadequate to operate the load and runs through the winter until the solar resource is once again sufficient.

Sizing of a PV–TEG hybrid system is governed by the operational modes and fuel-consumption requirements of the installation. The TEG is typically sized to handle the entire site load with a small excess for battery charging. The PV array and battery are sized in order to support the load for a certain fraction of the year, this fraction being based on a target for annual fuel consumption. The PV array will grow dramatically if it needs to carry the load during the winter months; design of cost-effective systems involves comparing delivered fuel and O&M costs with the capital costs of the array and battery.

Genset hybrid elements

Gensets of either the diesel or spark-ignition types (propane or gasoline) are used in many hybrid applications. The biggest advantages of engines are:

- They are on-demand power sources.
- They have relatively low capital costs.
- They are readily available from multiple vendors with well developed service and maintenance capabilities.

The disadvantages of engines are:

- They require regular maintenance and fuelling by trained personnel.
- They increase system complexity.
- They are noisy and they pollute.

Most gensets require an alternator, controller and rectifier in order to convert the rotational energy of the engine into DC power for battery charging. There are new DC output gensets that eliminate the rectifier by providing a regulated DC voltage directly from the alternator.

The choice of engine type must take into account the needs of the owner and operator as well as the conditions at the site. Diesel gensets are usually heavier and initially more expensive than spark ignition gensets, but are more efficient. In some cases, certain types of fuel may be easier to procure than others and will dictate the choice of genset type.

Starting a cold genset can be difficult and causes much wear. Starting an engine that is already near its operating temperature causes wear equivalent to one to four minutes of operating time;[1] starting an engine that has had time to cool down causes yet more wear, and repeatedly starting a genset that is near or below 0°C is quite damaging. The problems associated with cold-temperature operation of a genset can be greatly diminished by placing the genset in a heated enclosure (see below). Apart from this, several features can improve the genset's starting:

- a larger, more robust starting battery with additional cold-cranking amperage;
- glow plugs, located in the manifold, which preheat the engine block; this will facilitate combustion in diesel engines;
- oil-sump heaters or water-jacket heaters, which preheat the block and ensure good lubrication when the engine first starts running.

Gensets are mature commercial components that have a high level of inherent reliability. However, they do not approach the reliability of the PV module, which has no moving parts, no maintenance and a typical life exceeding 20 years.

Fuel considerations

TEGs run on natural gas or propane. Natural gas, which is typically piped directly to the point-of-use, is not usually considered for remote sites unless there is a pipeline already at the site. Propane, on the other hand, is a gaseous fuel that is readily available in pressurized tanks from numerous vendors.

Gensets can run on propane or a number of liquid fuels, including gasoline, diesel, kerosene and special low-temperature fuels such as 'JP4', 'JP8' and 'JA' (which can be considered as the same generic fuel type with different additives to improve flow at low temperatures).

The most significant drawback to propane in cold-climate applications is that it becomes liquid at temperatures below –35°C. Liquid propane cannot be used by gensets or TEGs, so a 'vaporizer' must be added. This consists of a nitrogen pressurizing system, which forces the liquid to flow, and a pre-heater element that reconverts the liquid to gas; this adds complexity, but is reliable when properly designed.

Recently 'regasified' propane has been found to contain trace amounts of liquid oils that are by-products of the manufacturing process. These trace amounts of liquid create problems for TEG burners with piezoelectric starting mechanisms. Regasified propane works without significant problems in spark-ignition engines.

The diesel fuels used in cold climates are often treated to improve their flow at low temperatures. Ignition of cold diesel fuel can be difficult, especially when the engine is just starting or warming up. In cold climates, preheating the fuel prior to combustion is recommended. This can be accomplished by drawing fuel from an intermediate storage tank located in a heated environment, such as the genset enclosure. The intermediate storage tank typically contains a day's supply of fuel.

Battery

Some large hybrid systems do not contain batteries, but the vast majority of PV–hybrid systems do. The battery has three functions and in some ways is the heart of the system:

- It is capable of providing surges of power, such as those required for starting motors, far greater than the maximum output of the generators.
- It transfers renewable energy from that point in time when it is available to the point in time when it is needed.
- It improves genset efficiency by permitting the genset to run at full load. Without a battery, the genset has to operate at the loading that provides the exact difference between the power demanded and the power available from the renewable sources. When this difference is small, the genset runs at partial loading, which dramatically lowers fuel efficiency. With a battery, the genset operates at full load and charges the battery; then the battery powers the load.

Balance-of-system components

Hybrid systems contain a number of components in addition to the array, second generator and battery:

- *Inverter.* For AC loads, an inverter will be required to convert DC electricity from the battery.
- *Chargers.* Most gensets and turbines generate AC; an AC-powered battery charger will be required.
- *Switches and fusing.* Hybrid systems generate dangerous voltages and currents. They are also complex systems, so that a person can be confused about which generators are on and which are off. Proper switching and fusing are essential.
- *Controllers.* A controller or controllers will be required to regulate charging, to determine when to start the generator, to carry out the genset ignition sequence and to control wind or hydro turbines.
- *Thermal enclosures.* Both gensets and batteries function poorly when cold; an insulated enclosure with a reliable thermal management system can be advantageous. TEGs and gensets produce waste heat that can be used to keep equipment warm. Heat losses can be minimized using insulated enclosures, but the system must not permit either the batteries or the genset to overheat. In addition, laws regarding ventilation and safety must be respected.

WHEN TO CONSIDER A PV–HYBRID SYSTEM

PV–hybrid systems typically fill the gap between small, stand-alone PV systems and prime-power gensets. While there are no rules determining whether a PV–hybrid system is the best choice, several considerations usually reveal whether it merits consideration.

PV–wind or PV–hydro hybrid systems should be considered whenever:

- Average loads are moderate or large (greater than, say, 30 W in a cold climate);
- *and* wintertime insolation is much lower than summertime insolation;
- *and* there is a reliable wind or hydro resource available during the winter;
- *and* the operating and maintenance requirements of the wind or hydro system can be satisfied at a reasonable cost.

PV–genset hybrid systems should be considered whenever

- Average loads are large (200 W or more at a mid-latitude site);

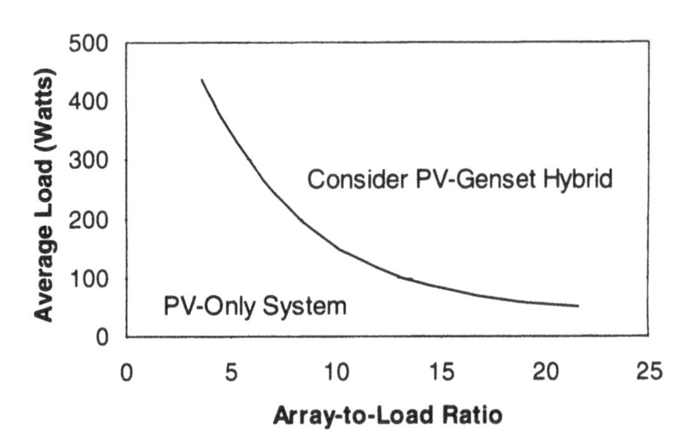

Figure 9.1 A rough indication of when PV–genset systems make sense (adapted from Sandia National Laboratories[2])

- *or* a PV-only system would have a very high array-to-load ratio.

Both of these conditions imply large PV arrays and large batteries and therefore very expensive PV-only systems. At some point, the high capital costs of a large PV system outweigh the additional complexity and cost of operating and maintaining a genset. The second condition further implies that a large array is partly necessitated by seasonal variation in the solar resource, and a genset operating during the winter would greatly reduce the size of the array. As a first approximation, this trade-off can be evaluated by sizing the PV-only system that would be required for the load (see Chapter 8), calculating the *array-to-load ratio* (divide the nominal array size, in watts, by the average load over the day, in watts) and then examining Figure 9.1. If this indicates that a PV–genset hybrid system might make sense, invest in a more accurate cost comparison.

The value placed on reliability and the costs of operation and maintenance will vary from installation to installation and will have great bearing on the comparison of a PV–genset hybrid and a PV-only system. If a site is very remote, then the costs of servicing and fuelling the genset will be very high and the curve in Figure 9.1 will shift away from the origin, i.e. larger arrays will be economical. In addition, concerns about the reliability of the genset, which contains many moving parts, may also shift the curve away from the origin.

DESIGN AND CONTROL OF PV–HYBRID SYSTEMS

The ultimate goal of the design and control of a PV–hybrid system should be to minimize the life-cycle cost of the system while satisfying constraints on reliability, availability of capital, operational and maintenance requirements and environmental impact. In practice, there are so many variables affecting the life-cycle cost that a complete analysis is difficult and rarely undertaken. Nevertheless, some basic considerations will lead to a reasonable, if not optimal, PV–genset hybrid system.

Design and sizing

There are a number of ways to approach the problem of designing and sizing a hybrid system.[3] In cold climates, the principal justification for a hybrid system is the seasonal variation in the sunshine; in a PV-only system, arrays and batteries would be large because they must be sized for winter sunshine availability. When this is the case, an obvious approach to hybrid-system sizing is to size the array and battery so that they meet a large fraction of the load during the summer, spring and autumn, and then size the genset so that it charges the battery effectively during the winter.

In some cases, the size of one or more components will be fixed and establish a constraint on the design. For example, PV-only systems that are being converted to hybrid systems will already have an array and the owner may want to use the existing array. Similarly, owners of gensets who want to add batteries and a PV array will not be interested in purchasing a different genset, and the rest of the system must be designed around this.

A third approach is to design the PV array and battery such that they will be able to satisfy a certain minimum load year round. This approach makes sense when a fraction of the total load is essential and must be powered at all times, even when the genset malfunctions.

Whichever approach is used, the design process should ideally involve careful analysis, simulation and comparison of costs. In reality, *ad hoc* design methods are more common, reflecting the fact that power on-demand makes it easy to find a design that works (though it may be hard to find the optimal design). The following considerations should guide component sizing, but do not supplant analysis and experience.

The battery

The battery can be sized in four steps:

1. Calculate the average load.

2. Decide on the type of battery to be used and determine how deeply this should be cycled. The optimal depth of discharge will vary from battery type to battery type; some should not be cycled deeply while others can be regularly discharged to a 20% state of charge. Keep in mind, however, that in hybrid systems cycling will limit battery lifetime and, the deeper a battery is cycled, the faster it ages. Also note that if temperatures are likely to be low enough to freeze the battery electrolyte, this will establish a lower limit on the depth of discharge. PV–hybrid systems in cold climates typically have a maximum depth of discharge of 50%.

3. Select the number of days of battery reserve; that is, if the PV array and genset were disconnected, the number of days a fully charged battery could operate the average load before reaching the maximum depth of discharge decided upon above. For PV–hybrid systems this is usually between one and five days, permitting the array to charge the battery during the day for night-time loads. A larger battery, though more costly, has a number of advantages:

   - It reduces the number of genset starts, which will reduce genset maintenance costs, especially in cold climates; this is especially advantageous at remote sites.
   - It provides additional battery capacity to compensate for that which will be lost due to poor battery performance at low temperatures.
   - It buffers day-to-day variation in the solar resource, resulting in more effective use of solar energy, especially if the *penetration ratio* (average power produced by the array divided by the average load) is high (approaching or exceeding one) in the summer. This reduces fuel consumption – which is of particular benefit at remote sites.
   - It makes control of the genset less critical, so it is easier to find reasonable control parameters.
   - It permits the same amount of reserve to be achieved with a higher minimum state of charge, or more reserve to be achieved with the same minimum state of charge, extending battery life.

4. Determine the size of the battery by multiplying the average load by the reserve (in hours) and dividing by the maximum depth of discharge. In general, hybrid systems have larger batteries in cold climates than in warm ones, because battery performance at low temperatures is poorer.

The genset

The genset can be sized on the basis of how long it will take to charge the battery from its maximum depth of discharge to a 90% state of charge. Conventionally the genset is sized such that it can do this in roughly five to eight hours (while powering the average load). Usually a larger genset is considered better: it costs little more; it charges the battery more quickly, runs less and thus needs less frequent maintenance;

and it facilitates expansion of the system in the future. There is an upper limit, however: the genset should not be sized so large that it supplies current higher than what the battery can accept – the C/5 current for the battery should not be exceeded (i.e. from a depth of discharge of 50%, it takes two hours to recharge to 90% state of charge). In cold climates, there are two additional reasons that larger gensets may not be advantageous. First, cold batteries accept charge reluctantly and inefficiently, so the charge current should be reduced. Second, during winter the genset may have to perform equalization charges, requiring the genset to operate at very low loading for several hours. Bigger gensets will waste more fuel during equalization.

The array
There is a great deal of flexibility in choosing the array size, as demonstrated by the approaches suggested above, since the genset will make up any shortfall. Nevertheless:

- The array should at least be large enough to justify the additional complexity of a hybrid system (compared to a genset/battery system); it should also be capable of performing an equalization charge during the summer – under full sun it should provide current exceeding the average daytime load plus the C/50 current of the battery.
- Larger arrays will lower fuel consumption, but eventually there will be diminishing marginal returns. For any proposed array size, examine the penetration ratio at midsummer and at the equinoxes. Penetration ratios approaching or exceeding one will result in wasted production, especially when the battery reserve is only one or two days. While this does raise the cost per kWh of useful PV electricity produced, it may be acceptable, especially at remote sites where minimizing fuel consumption has large paybacks.
- The array should not be so large that it produces currents exceeding the load plus the C/5 current of the battery, since the battery will be unable to accept this charge. If the array produces such high currents, but otherwise seems appropriate, consider enlarging the battery.

Control

In most PV–genset hybrid systems, four aspects of system operation must be controlled. First, charging of the battery must be regulated. Second, some dispatch strategy must determine when to turn on the genset, how hard to run it and when to turn it off. Third, the temperature of the genset must be kept within a reasonable operating range. Fourth, overdischarge of the battery must be prevented, through the use of a low-voltage disconnect, in those rare cases when the genset fails.

Charge control and dispatch strategy have three objectives: minimizing fuel use through maximizing the use of available renewable energy; minimizing genset wear by minimizing cold starts and run time; and maximizing battery lifetime. Unfortunately, these objectives are not fully mutually compatible. For example, minimizing fuel use may involve cycling the battery more deeply, which does not help to maximize battery lifetime. Thus, finding optimal control parameters is important.[4]

Starting the genset
The genset should be turned on under five conditions:

- The maximum depth-of-discharge has been reached.
- The temperature of the genset enclosure has fallen below a certain threshold, typically 0°C, and the genset must be started to warm the enclosure.
- The genset must be exercised since it has not been operated for a period of two to four weeks. This may be an important condition in cold-climate hybrid systems, since during the summer sunshine will be abundant and may be able to power the load entirely.
- The battery has not been equalized in the last two to four weeks; note that more frequent equalization results in faster water loss from the batteries.
- A very large load is being operated for a sustained period.

Genset loading
In most PV–hybrid systems, it makes sense to run the genset at full load (i.e. 80% of the nameplate capacity of the genset) whenever the battery is able to accept the resulting current. There may be exceptions at those rare sites where fuel is very cheap and the genset is not much larger than the average load.

Stopping the genset
There are a number of approaches of varying sophistication.[5] These can be illustrated by imagining how the stopping strategy might have evolved at a particular site as an operator optimized a hypothetical system.

Originally, the operator programmed the controller to stop the genset when the battery was fully charged. Unfortunately, because a battery can accept less charge as it approaches 100% state of charge, this resulted in the genset running for several hours at partial loading at the end of every charge. This was very inefficient.

To reduce fuel consumption, the operator reprogrammed the controller so that the genset stopped when the battery reached a 90% state of charge. This meant that the genset could operate at full load all the time. But soon the operator noticed that the capacity of the batteries seemed to be declining. This was due to stratification of the electrolyte in the flooded batteries, diverging cell voltages and sulphation resulting from cycling the battery between two partial states of charge.

The operator changed the strategy so that if the genset was running and equalization had not occurred in the previous three weeks, it did an equalization charge. This solved the battery problems.

The next spring, however, the operator noticed that after the genset charged the battery to 90% state of charge and shut off, the electricity from the array would be wasted because the battery was already fully charged. This was not usually a problem during winter, since the penetration ratio was low, but during the rest of the year, when penetration ratios approached or exceeded one, much solar energy was being wasted. This meant unnecessarily high fuel consumption.

In response, the operator started adjusting the stopping strategy on the basis of the penetration ratio for the season. During winter, when the penetration ratio was below 0.25, the genset would turn off at 90 % state of charge. This minimized cold starts and genset run time. At midsummer, when the penetration ratio exceeded one, the genset would raise the state of charge by only 20% of the maximum depth of discharge (i.e. for a maximum depth of discharge of 50%, the genset would raise the state of charge to 60%). For the rest of the year, when the penetration ratio varied between 0.25 and 1, the operator varied the stopping criterion between the two extremes.

As a final optimization, the operator adjusted the frequency of equalization. During summer, the controller performed equalization every two weeks. Since the array was usually capable of equalization, this increased genset run time very little. During winter, equalization occurred only once every four weeks. This, the operator judged, was a reasonable compromise between the temporary reduction in battery capacity due to stratification and the increased fuel usage for equalization by the genset.

Under different operating conditions, different strategies might work better. Finding the best strategy often requires simulation and careful analysis. This can be a worthwhile investment, since it can greatly influence fuel consumption and maintenance requirements.

Optimizing design and control

In the above sections, design and control have been treated separately, as if system sizing and selection of the control strategy did not affect each other. Although this is often done, it ignores the fact that a system is more than its components. By optimizing each component individually, one risks missing the optimal configuration of the entire system. This can be found only by studying the interactions of the various components under various control strategies.

As one example, consider this scenario. In order to minimize fuel consumption, a designer decides to make maximum use of available solar energy by choosing a large battery. Recognizing that a full battery will be unable to make use of solar energy, he sets the control strategy so that the genset turns off when the battery reaches 60% state of charge. Unfortunately, the designer did not examine how the whole system would function together. It turns out that even in midsummer the penetration ratio is about 0.7; this means that the battery has a tendency to run down as opposed to charge up. As a result, the genset starts and stops frequently, causing much wear. Furthermore, the genset must perform equalization since the array is usually incapable of this. This consumes a lot of fuel. Eventually the designer adjusts the controller so that the genset shuts off only when the batteries reach 90% state of charge. This reduces genset cold starts and sometimes the array can equalize the battery. The system operates better, but the expensive battery was largely a wasted purchase.

REFERENCES

1. Bleijs JAM, Nightingale CJE, Infield, DG (1993). Wear implications of intermittent diesel operation in wind/diesel systems. *Wind Engineering*, 17, pp.206–219.
2. Sandia National Laboratories (1995). *Stand-alone Photovoltaic Systems: A Handbook of Recommended Design Practices.* Albuquerque, NM: Sandia National Laboratories.
3. CANMET (1991). *Photovoltaic Systems Design Manual.* Ottawa: Natural Resources Canada.
4. Usher E, Ross M (1998). Recommended Practices for Charge Controllers (International Energy Agency PVPS Task III Report). Varennes, Québec: Natural Resources Canada.
5. Barley C, Winn CB (1997). Remote hybrid power systems. *Advances in Solar Energy*, 11, pp.109–158.

10 Seasonal storage for photovoltaic systems

Juha Vanhanen
(Helsinki University of Technology)

Stand-alone power applications are a significant market sector for PV power systems today and into the near future. An important technical constraint for wider use of PV power systems in stand-alone applications in cold climates is the difficulty of storing summertime energy for wintertime use. At higher latitudes, a solar system has surplus production during the summer and an energy deficit during the winter. At very high latitudes, there is practically no solar energy available during the wintertime and therefore some kind of seasonal storage is needed for applications which operate year-round. The winter energy demand of a solar-based system can be met by either a back-up primary battery or genset (see Chapter 9), or the system can be designed to be completely self-sufficient, i.e. able to store excess summer PV production for the wintertime.

PV–BATTERY SYSTEMS

PV–battery systems use either primary or secondary batteries to meet the load during periods of low solar irradiation. Primary batteries cannot be charged and therefore they must be changed when they are completely discharged. Secondary batteries can be recharged, but typical high-quality batteries must be replaced after 5 to 15 years of PV system operation (see Chapter 6).

The primary battery typically serves as a back-up to a conventional stand-alone PV system with secondary batteries. For most of the year, the PV array charges the secondary batteries, which power the load at night and on the occasional cloudy day. During winter and other periods of low solar power, the secondary batteries are unable to supply the load and a controller switches over to the primary battery.

PV–primary battery systems are generally feasible only for small-scale applications because of the limited energy density of batteries. Once activated, primary batteries have a maximum lifetime of one to three years; the batteries must be replaced at least as frequently as once every couple of years. Because of the weight of the batteries, transporting batteries to and from a remote site can be extremely expensive. Furthermore, disposal of the used batteries represents a significant environmental cost, especially when considered over the lifetime of the system.

PV–primary battery systems are typically found in Arctic regions, where rime is a concern, or where primary batteries were used traditionally but now photovoltaics have been introduced. For example, many low-power telecommunication repeater stations in Western Canada were powered in the past by primary zinc–air batteries. Over the past ten years, these have been converted to PV systems designed to work without seasonal storage. When the PV system is installed, the existing zinc–air batteries are left in place to back-up the PV system. Since the battery self-discharges with time, the primary battery provides only a temporary back-up, but at least it prevents loss of load due to design flaws or equipment failures during the first year of operation.

Systems using secondary batteries for seasonal storage appear, at first glance, to be no different from regular stand-alone PV systems: they have all the same components and operate similarly. The key difference is the sizing of the PV array. In regular stand-alone PV systems, the array is sized to provide sufficient power to the load during the least sunny month of the year; thus the battery is used only to buffer daily or weekly shortfalls in production. In contrast, with seasonal storage, the array is deliberately undersized: it is incapable of providing sufficient power to the load during the least sunny month or months of the year and thus the battery must make up the difference. This approach makes sense when the solar resource during winter is so minimal that conventional sizing of the array is prohibitively expensive or impossible: for example, above the Arctic Circle, polar night makes it impossible for the PV array to power the load and seasonal storage is necessary (see Chapter 8).

Seasonal storage using secondary batteries is feasible but expensive. In order to store sufficient energy for the winter, enormous battery banks are generally required, especially since the cold reduces the usable capacity of the batteries. In Arctic locations, low temperatures during autumn and spring may limit the battery's ability to accept charge, requiring yet larger battery banks. A large array is required to ensure that charge currents will be sufficiently high, relative to the size of the battery bank, to achieve full charge. When kept at a partial state of charge for a long period of time, lead–acid batteries tend to sulphate and the state of charge of the different cells diverges. This may cause accelerated deterioration of the battery and thus require more frequent battery replacement. On the other hand, self-discharge of secondary batteries proceeds very slowly at cold temperatures and this is not generally an impediment to seasonal storage in cold climates.

HYDROGEN-BASED STORAGE SYSTEMS

The high cost of the large battery banks required for seasonal storage and the considerable expense of maintaining and refuelling a genset at remote locations have stimulated the search for novel solutions to seasonal energy storage and back-up. Increasing attention has been paid to hydrogen-based systems, which seem to be interesting and promising for the long run.

Recent developments in hydrogen technology open new possibilities for solving the problems related to seasonal energy storage. Hydrogen is a clean, sustainable, transportable and storable fuel having the high energy density of 39 kWh/kg (not including the weight of storage vessels), which is about four times higher than that of gasoline. It can be converted to electricity in a fuel cell with very high efficiency without any of the NO_x emissions associated with conventional combustion reactions. In the case of solar electricity, excess summer photovoltaic electricity production may be converted to hydrogen to be used in wintertime. Thus, hydrogen would be an excellent energy carrier to level the seasonal variation of irradiation.

The key components of the hydrogen-based energy storage system are an electrolyser, a fuel cell and a hydrogen store. In the electrolyser, water is decomposed into hydrogen and oxygen by electrical energy. The most common types of electrolysers are alkaline electrolysers and solid-polymer electrolysers. The alkaline electrolyser has a potassium hydroxide solution as electrolyte, i.e. hydroxide ions are used as charge carriers in the electrolyte, while the solid-polymer electrolyser has a proton-conducting polymer membrane to transfer hydrogen ions from the anode to the cathode. The efficiencies of state-of-the-art electrolysers in converting from electricity to hydrogen are about 80%.

For hydrogen storage, a conventional pressurized steel vessel is the cheapest option, but its energy density is quite modest. Especially in mobile applications, higher energy densities are needed. Therefore, new options for hydrogen storage, such as metal hydrides and light-weight composite containers, have been suggested. The former increase the energy density per volume while the latter improve the energy density per mass. State-of-the-art metal hydrides may reach the energy density per volume of 3 kWh/litre; carbon–graphite composites may reach nine times the energy density per mass of steel vessels. However, both metal hydrides and composites are still expensive compared to the conventional 200 bar steel bottles.

The most crucial component in hydrogen-based PV systems is the fuel cell. Fuel cells, in general, have a very wide power range and they match extremely well with small power needs. The conversion efficiency from fuel to electricity in a fuel cell is some 50%, which is several times higher than a typical genset. Furthermore, no emissions and noise are produced during operation, unlike fossil-fuel back-up generators. There are several types of fuel cells and usually they are categorized according to the electrolyte used, i.e. alkaline, phosphoric-acid, solid-polymer, molten-carbonate and solid-oxide fuel cells. The phosphoric-acid fuel cells are relatively close to commercial penetration of the market – presently there are several 200 kW fuel-cell units manufactured by ONSI Corporation operating all over the world. Furthermore, a recent initiative by Ballard Power Systems, Ford Motor Company and Daimler–Benz to commercialize solid-polymer fuel cells has the potential to reduce the cost of this fuel cell technology.

PV–fuel-cell system

It has been suggested that a novel PV–battery system backed-up by a fuel cell – called a PV–fuel-cell system – could eliminate the need for seasonal battery storage in applications that operate year-round.

The PV–fuel-cell system consists of a PV array, a battery, a fuel cell, a hydrogen store and a control unit (see Figure 10.1). In this system, the PV array produces electricity for the load when solar radiation is sufficient. The battery is used as short-term energy storage to buffer short-term fluctuations in solar radiation and also to store the daytime electricity production for night-time use. The fuel cell is used as a back-up generator to produce electricity when the state of charge of the battery is low. In this system, hydrogen is replenished periodically, e.g. once a year. Conceptually, it is similar to a PV–genset hybrid system, but with the genset and its fuel tank replaced by a fuel cell and its fuel supply.

The PV–fuel-cell system has several advantages compared to the conventional PV–battery system. As the PV–fuel-cell system has a back-up power generator, the PV array and battery do not need to be oversized, which is often the case with the PV–battery systems. A promising future option is to use methanol as a fuel. In this case, a methanol reformer is needed to produce hydrogen for the fuel cell, but transporting and storage of the fuel are easier because methanol is a liquid fuel.

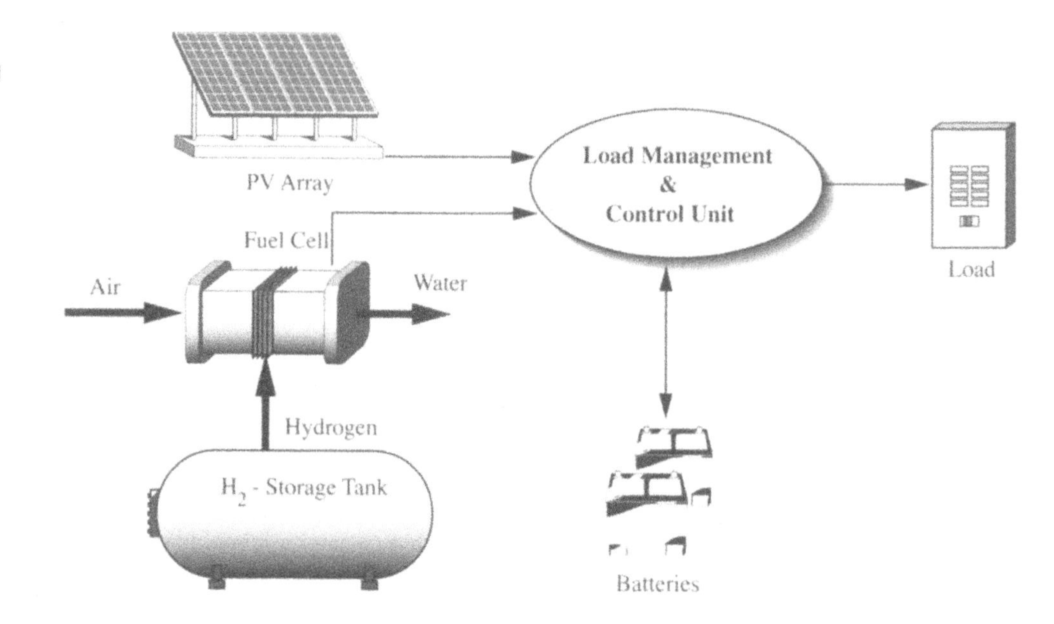

Figure 10.1. Schematic diagram of the PV–fuel-cell system

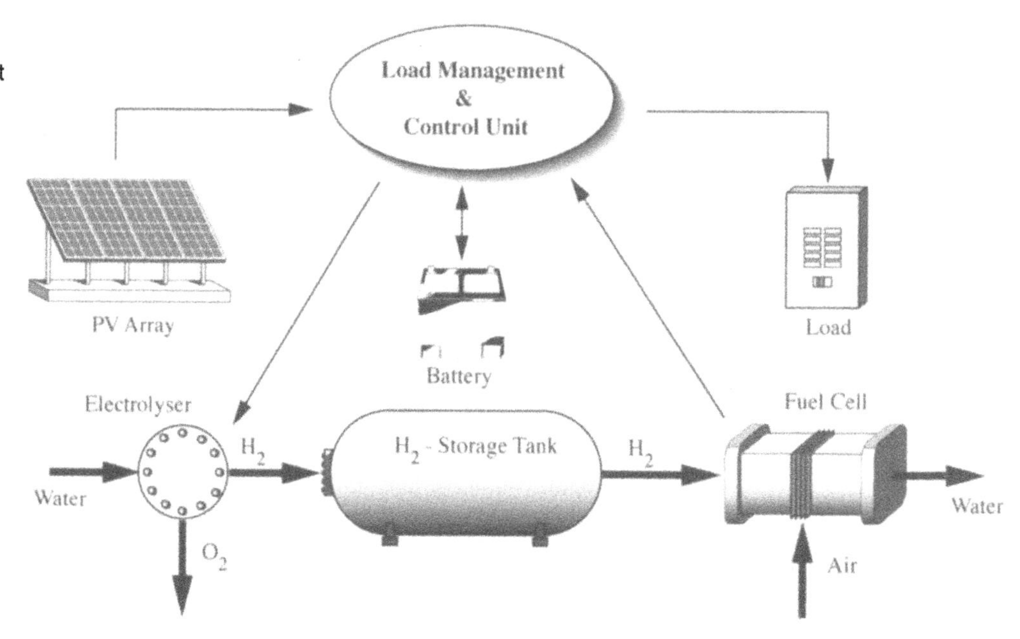

Figure 10.2. Schematic diagram of the self-sufficient PV–H_2 system

Self-sufficient PV–H_2 system

Self-sufficient PV–H_2 systems have been envisaged as replacements for PV–genset hybrid systems. They would be very attractive for remote power-supply systems that are located far away from the electric grid and require complete self-sufficiency owing to difficult access for servicing.

The self-sufficient PV–H_2 system consists of a PV array, a battery, an electrolyser, a hydrogen storage tank, a fuel cell and a control unit (see Figure 10.2). In this system, the load is met directly by PV electricity whenever possible. During periods of low irradiation, the load has to be met by either short- or long-term energy storage. The short-term energy storage is carried out by a battery, which is used as a buffer between the load and the PV array to level short-term fluctuations in irradiation, as well as to store the daytime electricity production for night-time use.

The long-term energy storage is carried out by the H_2-based energy storage system, comprising an electrolyser, a

Table 10.1. Comparison of the seasonal storage options for PV systems

	PV–battery system	PV–fuel-cell system	Self-sufficient PV–H_2 system
Maintenance	Primary battery replacement, secondary battery watering	Replenishment of hydrogen	Almost maintenance free (currently once a year)
Costs	High investment costs for secondary batteries, high replacement costs for primary batteries	Operational cost dominates	Investment cost dominates
Reliability	Battery is critical component	Fuel cell is critical component	Electrolyser and fuel cell are the most critical components
Size	Compact	Compact	Depends on the pressure of the storage
Environmental aspects	Requires recycling of batteries	Environmentally friendly	Environmentally friendly
Applications	Small applications < 1 kW	Small applications < 1 kW	Applications located far away from services
Status	Proven	Technical feasibility trials	Technical feasibility trials

hydrogen storage tank and a fuel cell. In this system, excess summer photovoltaic electricity is fed to the electrolyser to produce hydrogen and oxygen from water. The hydrogen produced is stored over the season to be converted back to electricity in the fuel cell in the winter. Thus, hydrogen is used as a storage medium to shift solar energy from summer to winter.

Market potential of hydrogen-based PV systems

The major challenge to be overcome in hydrogen technology is the cost reduction of the electrolysers and fuel cells. According to Barbir and Gomez,[1] the cost of a prototype solid polymer fuel cell is today about 3000 USD/kW. In order to be competitive with existing technologies, the cost of fuel cells should be reduced below 1000 USD/kW for stationary and below 100 USD/kW for automotive applications. Cost reductions can be achieved by reducing catalyst loading of electrodes, by increasing the lifetime of the cells and by developing more efficient manufacturing processes.

Even though the costs of the hydrogen energy components are still high, there are market niches for hydrogen-based PV systems even today. Self-sufficient PV–H_2 systems are especially feasible in remote locations far away from services. In fact, the lower maintenance cost due to complete self-sufficiency is the major factor that may change the economics in favour of the self-sufficient PV–H_2 system. Therefore, the first commercial PV–H_2 applications will be autonomous systems, e.g. telecommunication links far away from services. If these special systems prove to be technically and economically feasible in practice, the next generation of PV–H_2 systems could be self-sufficient power-supply systems for isolated islands and villages.

The PV–fuel-cell system may have huge market opportunities in numerous small-scale applications, if the fuel cells can utilize methanol as a fuel either directly or indirectly via a reformer. Two large car manufacturers, Daimler–Benz and Toyota, have introduced methanol-based fuel-cell cars that have methanol reformers and solid-polymer fuel cells. If these automotive systems and cost-reduction targets prove to be feasible in practice, they will encourage the use of compact PV–fuel-cell systems for stationary applications.

COMPARISON OF SEASONAL STORAGE SYSTEMS

The properties of the PV–battery, PV–fuel-cell and self-sufficient PV–H_2 systems are compared in Table 10.1. The PV–battery system is compact and reliable, but it is environmentally unfavourable because the battery bank must be changed periodically – in the case of the primary battery after every discharge. Furthermore, this system is feasible only for small applications because of the limited energy density of the batteries. The PV–fuel-cell system is a compact and environmental friendly system for slightly larger applications. In this system, operating cost dominates because replenishment of hydrogen fuel is needed periodically. The self-sufficient PV–H_2 system is almost maintenance-free (current designs require maintenance once a year) and thus it is especially attractive for remote applications. As a result of the high prices of electrolysers and fuel cells, its investment cost is high.

REFERENCE

1. Barbir F, Gomez T (1997). Efficiency and economics of proton exchange membrane (PEM) fuel cells. *International Journal of Hydrogen Energy*, 22, pp.1027–1037.

FURTHER READING

Billings RE, Sanchez M (1995). Solid polymer fuel cells: An alternative to batteries in electric vehicles – an overview. *International Journal of Hydrogen Energy*, 20, pp.521–529 (1995).

Blomen LJMJ, Mugerwa MN (1993). *Fuel Cell Systems*. New York: Plenum Press.

Bockris, JO'M, Srinivasan S (1969). *Fuel Cells: Their Electrochemistry*. New York: McGraw-Hill.

Garcia-Conde AG, Rosa F (1993). Solar hydrogen production: A Spanish experience. *International Journal of Hydrogen Energy*, 18, pp.995–1000.

Grasse W, Oster F, Aba-Oud H (1992). HYSOLAR: the German–Saudi Arabian program on solar hydrogen – 5 years of experience. *International Journal of Hydrogen Energy*, 17, pp.1–7.

Haas A, Luther J, Trieb F (1991). Hydrogen energy storage for an autonomous renewable energy system – Analysis of experimental results. *Proceedings of the ISES 1991 Solar World Congress*, Denver, pp. 723–728. Oxford, UK: Pergamon Press.

Hollenberg JW, Chien EN, Lakeram K, Modroukas D (1995). Development of a photovoltaic energy conversion system with hydrogen energy storage. *International Journal of Hydrogen Energy*, 20, pp.239–243.

Kauranen PS, Lund PD, Vanhanen JP (1994). Development of a self-sufficient solar–hydrogen energy system. *International Journal of Hydrogen Energy*, 19, pp.99–106.

Ledjeff K, Heinzel A, Peinecke V (1994). Development of pressure electrolyser and fuel cell with polymer electrolyte. *International Journal of Hydrogen Energy*, 19, pp.453–455.

Lehman PA, Chamberlin CE (1991). Design of a photovoltaic–hydrogen–fuel cell energy system. *International Journal of Hydrogen Energy*, 16, pp.349–352.

Linden D (1984). *Handbook of Batteries & Fuel Cells*. New York: McGraw-Hill.

Odgen JM, Williams RH (1989). *Solar Hydrogen: Moving Beyond Fossil Fuels*. Washington, DC: World Resource Institute.

Prater KB (1994). Polymer electrolyte fuel cells: a review of recent developments. *Journal of Power Sources*, 51, pp.129–144.

Vanhanen JP, Kauranen PS, Lund PD, Manninen LM (1994). Simulation of solar hydrogen energy systems. *Solar Energy*, 53, pp.267–278.

Vanhanen JP, Lund PD (1995). Computational approaches for improving seasonal storage systems based on hydrogen technologies. *International Journal of Hydrogen Energy*, 20, pp.575–585.

Veziroglu TN, Barbir F (1992). Hydrogen: the wonder fuel. *International Journal of Hydrogen Energy*, 17, pp.391–404.

Winter C-J, Nitsch J (1988). *Hydrogen as an Energy Carrier*. Frankfurt, Germany: Springer-Verlag.

11 System modelling and simulation

Bengt Perers
(Vattenfall Utveckling AB)

The complex interactions that occur among the components of a PV system are not accounted for intuitively, especially since they are driven by constantly varying conditions of sunlight and temperature. Computerized models can reveal much about the behaviour of real systems, permitting the designer to optimize and test a system before it is actually built.

INTRODUCTION TO MODELLING AND SIMULATION

A researcher or system designer who wants to know how a particular PV system will behave under a given set of conditions faces a choice: he or she can either build and experimentally test the system, or mathematically describe the pertinent aspects of the system and then use this mathematical model to investigate system operation. The latter approach is much cheaper and requires less time.

Mathematical models of PV systems require that the user supply component parameters and weather data. Component parameters, e.g. array size and battery capacity, specify the nature of the constituent components. Weather data include the environmental conditions, such as the availability of sunlight, the ambient air temperature and the wind speed.

Normally a mathematical model will be incorporated into a computer program and 'fed' a time series of weather data. This is simulation, or the study of system behaviour via a computerized mathematical model whose inputs are controlled so that they reflect real-world conditions. Simulation is helpful for dynamic systems, where the behaviour at a given time is not entirely determined by the values of the inputs at that time. For example, battery state of charge in a PV system is related to past as well as present levels of sunshine; by simulating a PV system over a year, the minimum state of charge can be estimated.

Once component models have been validated against measurements in real outdoor operations, simulation can replace expensive experiments. Many variations can be studied to find the optimum configuration of a system for a given site, new ideas can be tested and the correct operation of an existing system can be verified through comparison with simulation results.

Simulation of PV systems for cold climates is special in two ways: first, one must ensure that the mathematical models are valid over the range of temperatures (and other environmental conditions) that will be encountered. Second, the seasonal variations in temperature and the availability of sunlight require investigation of the system operation over the whole year. This complicates experimental investigations and thus makes simulation even more valuable.

TYPES OF SIMULATION AND MODELLING TOOLS

Three different types of tools exist for PV system calculations: simulation tools, sizing tools and nomograms.

Simulation tools

These are normally used to analyse the performance of a system as specified by the user. Such tools typically require that the user supply both a large number of component parameters and, for each hour of the simulation period, temperature and insolation data. The combination of detailed component models and hourly weather data results in a powerful and potentially very accurate tool. The flexibility of many simulation tools also permits them to investigate a wide range of system configurations. In return, the simulation software demands more knowledge and diligence on the part of the user; incorrect input parameters can lead to gross errors. Paradoxically, *the more powerful and flexible the tool, the greater the potential for error and inaccuracy.*

Because simulation demands much from the user, it is generally employed by researchers or experienced designers working on large or very important systems. Simulation is very useful for investigating specific hypotheses concerning system behaviour. For example, comparing monitored temperatures, currents and voltages with simulation predictions can reveal whether an existing system is operating correctly. It is also helpful, when the operation of a system during a short period of time (say, a day) is of concern, for identifying extremes in system behaviour (e.g. lowest battery voltage or highest module temperature) and for accurately determining average energy flows.

The rapid development of personal computers has now reduced calculation times to the point that a detailed simulation for a period of a year or more may take only seconds or minutes. As a result, many different variations can be investigated in a short time, facilitating system sizing. Sizing involves repeatedly running the model and manually searching for the best combination of component parameters and sizes. A certain amount of expertise is necessary and simulation does not replace sizing software for most users.

Simulation software must be capable of presenting results in a variety of ways so that the user can evaluate the results and be confident that the simulation runs correctly. Many tools present a synopsis of the simulation results first, so that the user can identify major errors, and subsequently supply details related to the problem under investigation.

Sizing tools

These tools determine the size of the components required to meet a specified load at a specified site. Simulation tools can do this, too, but require tens or hundreds of simulations to converge on the optimal configuration; the additional cost in terms of the time required to learn and use the simulation tool is justified only for special or expensive systems. Thus, sizing tools are aimed at routine applications and users who are interested in a quick and easy answer. Because of this, they often rely on simplified component models that free the user from the burden of specifying large numbers of input parameters and require insolation and temperature data on a monthly, as opposed to hourly, basis. These simplifications tend to reduce accuracy, but leave less room for human error and make sizing software more user-friendly.

Intended for less exacting users, sizing software often focus on presenting output in a straightforward manner. Many programs display results graphically. In addition, programs often include routines for costing and economic analysis, and some attempt to perform life-cycle costing. While there is no reason that simulation tools cannot include these features, they are more common in sizing tools.

Some PV component manufacturers distribute their own sizing tools. These tools are designed to work with the products produced by the manufacturer and thus require fewer input parameters. Corporate backing usually ensures a nearly bug-free tool with a worldwide weather database. The user should be aware that these tools may be designed to oversize those components sold by the manufacturer.

Nomograms

Nomograms are clever graphical constructions that illustrate the relationship between two or more key variables and permit the rapid but approximate solution of specific design problems. (This definition reflects the usage of the term in solar-energy circles; 'nomogram' has a more restrictive definition in mathematics.) Although they cannot replace simulation and sizing tools, nomograms are helpful for first estimates, quick checks and the development of a deeper understanding of the relationships between variables. One common nomogram is a chart showing the position of the sun in the sky at any time during the year. With a few simple calculations, a user can superimpose the profile of a tree, building or other obstacle, immediately revealing how much the obstacle will shade the array.[1] The solution is somewhat imprecise, but the method is much quicker and easier than calculating the shading of the array using equations. Furthermore, the graphical solution makes interpretation of the results obvious.

In recent years, nomograms have fallen out of favour, presumably because of the availability of powerful computerized sizing and simulation tools. This is unfortunate, because nomograms can help understand many aspects of system operation. The curves of a nomogram graphically illustrate trends and tendencies; a single run of a simulation or sizing tool corresponds to a single point on a nomogram. Rare is the person who can make generalizations on the basis of a few simulation runs, dependent on dozens of variables; in contrast most people grasp the implications of nomogram. Despite their simplicity, nomograms are sufficiently accurate for many purposes: constructed by carefully selecting the axes, holding less significant variables constant, and condensing large numbers of simulation results, they reflect the insight of their developers.

USING SIZING AND SIMULATION SOFTWARE

Why should one use these difficult-to-learn, time-consuming and expensive tools when simple hand-calculation tools can yield results? The answer is that they are cost-effective in the long run. Errors in the design of a photovoltaic system result in either undersizing or oversizing. The former leads to system failures that may be very costly to remedy, especially for remote stand-alone PV systems; the latter results in systems that cost more than they need to.

Sizing and simulation tools are not always used. Some designers maintain that their experience and intuition are sufficiently accurate and as evidence they point to previous designs that have worked reliably. In some cases this may be true, but a reliable system is not necessarily well-designed – it may simply be oversized. In one study of 22 stand-alone PV systems for developing countries, for example, monitoring demonstrated that all of the five systems designed using computerized simulation tools were properly dimensioned.

In contrast, of the 17 systems designed using other procedures, four were undersized and seven were oversized. Of the oversized systems, most furnished two to four times too much power (Gordon,[2] citing Ratajczak[3]). This study is somewhat dated now, but there is evidence to suggest that the situation has changed little.

In cold climates, intuition is yet more prone to error, since the paucity of solar energy during the winter leaves little margin for error. Unfamiliar climates and systems can throw off even experienced PV designers; most designers apply conservative margins of safety. Simulation and sizing tools can help identify problems and pare down safety margins.

Simulation and sizing tools are essential, but alone they are insufficient. To be used successfully, they must be married to good judgement, diligence, scepticism and understanding. It is the responsibility of the user to do the following:

- Ensure that the models used in the simulation are valid for the range of conditions being simulated.
- Assess the limits on the accuracy of the simulation; digital simulations are highly precise (loss-of-load probabilities are sometimes expressed to one or two decimal places), but are not always accurate.
- Be aware of real-world considerations not reflected in the simulation, such as snow, ice and dust accumulation on the array, shading and battery ageing.
- Provide component parameters that accurately reflect the particulars of the system.
- Acknowledge the natural variation inherent in weather data, i.e. there are no limits to weather extremes.
- Account for inaccuracies in input weather data, especially data that has been interpolated, transformed or synthesized.
- Double-check software output to ensure that, in the complex interactions between the user, the input data and the models, something has not gone awry.

In certain circumstances, existing simulation and software tools may be unacceptable. The user should inform software developers of this, so that future releases can reflect the real needs of their users.

SELECTING SOFTWARE

Many sizing and simulation tools are available.[4] Selection of a software package should reflect a number of considerations. The first of these consideration is *not* price. Instead, one should examine the total time it will take to learn and to use the program, as well as other factors that will influence the future usefulness and cost of using the program. Assess future as well as present needs and *choose software that will be useful for a long time*; learning new software thoroughly is time-consuming.

In the age of fast personal computers, older programs that were designed to minimize calculation time are not necessarily appropriate. A sizing program may nowadays use the same kind of detailed models as simulation software.

Some questions that should be asked before selecting a sizing or simulation package relate to methods, capabilities, support, imput data, models and features.

Method selection

- What kind of tool, i.e. sizing or simulation, is needed?
- What accuracy is needed?
- How much time can be spent learning the software? (This time is often much more costly than the software itself.)

Basic capabilities

- What system types are covered? What types need to be included, now and in the future?
- What output data is needed and is it available from the program? How is output displayed?
- How easy is it to use the software?
- How easy is it to learn the software?

Support

- What user support is given and what is the cost?
- Is there a group or company behind the software that continuously updates the software to reflect new options, components and applications?
- How widely used is the software? (A large pool of users should result in a reduced number of bugs.)

Climate data and radiation models

- What locations are covered in the weather data available to the program?
- How easy is it to add new locations to the weather database and what is the cost?
- What is the format for weather data, and is weather data commonly available in this format?
- Has the program been validated against monitored data from a cold-climate site? Was the accuracy acceptable?
- Does the radiation model include shading, module tilt, azimuth, tracking and reflectors?

Models

- Are the models well documented or are they treated as black boxes? Is this important to the user?

Features

- If it is simulation software, can it perform optimization or sensitivity analyses by iteration?
- Can the software perform an economic analysis based on the system performance calculations?

PARTICULAR CONCERNS FOR MODELLING IN COLD CLIMATES

Simulation and sizing software is not always developed with cold regions in mind and, as result, programs sometimes fail to reflect the true behaviour of components and systems in such areas. Cold climates are extreme climates and extreme behaviour is often hard to model. No software yet in existence works perfectly for cold regions; the user should assess software packages in light of the following.

Period of simulation

Many programs simulate system operation over the period of one year, starting in January and ending in December. This can be problematic for simulations of cold-climate stand-alone systems. In such systems batteries are often sufficiently large that they buffer not only day-to-day variation in the solar resource, but also transfer some energy from late autumn into the winter. During the winter, the batteries typically undergo a deep discharge, from which it takes weeks or months to recover. Thus, these simulations start and end precisely at the most critical time of the year and can give misleading results. For example, a simulation starting in January and assuming a state of charge of 100% may find that a system operates acceptably through to December. If the state of charge at the end of December is 60%, however, running the simulation for an additional year might reveal that the system would fail in the second January. In short, the first year just served to 'prime' the model and establish the correct initial conditions.

This type of problem can be addressed in three ways. First, some simulation programs let the user specify the start date; by setting this to some time in midsummer, the user ensures that the model will be primed for the following winter. Second, some simulations can be run for a period of two years. Third, the user can run the simulation once, check the state-of-charge at the end of the first year, and then enter this as the initial state-of-charge and run the simulation again. Programs that do not permit at least one of these three options are not appropriate for simulating stand-alone systems in cold climates.

Modelling battery operation

Of all the components in a photovoltaic system, the battery is the most difficult to model accurately. This is especially true in cold climates, since battery operation is greatly affected by operating temperature.

There are a number of different approaches to battery modelling. Perhaps the most common among simulation and sizing software is the 'energy transfer' model,[5] which treats the battery as a 'black box' store of energy, and tallies charge into and out of the battery. This approach has three weaknesses for cold-climate applications:

- At lower temperatures, less charge can be drawn from a battery, since the electrolyte may freeze (in lead–acid batteries) and ions move sluggishly at low temperatures, hampering charge transport at higher discharge rates. Ensure that the battery capacity is adjusted to account for *both* electrolyte freezing and the charge transport at a representative discharge rate.
- At low temperatures, it is difficult to fully charge a battery; in real life, the controller end-of-charge voltage set points must be raised. Models which ignore this will make use of charge current that a real battery would not be able to accept.
- Because the resistance of a battery increases whenever the battery is cold or nearly fully charged, voltage efficiency and energy efficiency decline at lower temperatures and higher states of charge. Often energy-transfer models neglect this and thus overestimate the energy liberated from the battery.

The energy-transfer model works for rough approximations, but because it cannot predict the battery voltage at different currents and states of charge, it cannot be used in simulations that attempt to model accurately the interactions between the PV array, the controller, the load and the battery. The model can be adapted to permit study of these interactions: battery voltage can be predicted from a family of curves of voltage versus state of charge, with curves taken at various temperatures and currents.[6] This addresses the three problems listed above, but the user must ensure that the family of charge and discharge curves cover the temperatures and currents that will occur during the simulation; extrapolating to colder temperatures and higher rates frequently leads to errors.

'Simulation models' also predict battery voltage.[5] These include:

- semi-empirical models, such as the 'Shepherd' model, which relate battery voltage to the underlying potential of the electrodes, the internal resistance of the battery, the electrolyte concentration and other factors;
- 'behavioural' models, which attempt to model the physical and electrochemical state of the battery;

- 'electrical' models, which postulate an equivalent electric circuit for the battery and then assume that the battery behaves as this circuit would.

The suitability of these models to cold-climate applications will depend on whether the large set of parameters used in these models are constants or treated as functions of temperature, with the functions defined over an appropriate temperature range. One way to evaluate whether any model is appropriate for cold climates is to examine how well it overcomes the three traditional weaknesses of energy-transfer models, listed above.

Ideally, battery models would recognize that a battery 'remembers' how it has been treated, both in the short term and in the long term. For example, in lead–acid batteries, an ideal model would account for stratification of the electrolyte, which can occur when cycling between partially full and partially empty, sulphation and various symptoms of aging. All of these affect the discharge capacity, efficiency and charge acceptance of the battery, and all are concerns in cold climates, where batteries spend most of the winter partially discharged. Most models ignore these, and understandably so; these are complex phenomena and attempts to treat them in non-specialized software packages would probably fail.

The complexity of the storage battery has prevented the development of highly accurate models. There is little that the software user can do about this, except to be aware of the limitations of the software.

Battery lifetime modelling

Conventional wisdom has it that one of two factors will limit the lifetime of a battery: grid corrosion or deep cycling. Grid corrosion is exponentially related to battery temperature, so decreasing the battery temperature by 10°C should double the lifetime of a battery that is not being deeply cycled. The number of deep cycles a battery will experience is loosely determined by the size of the battery in comparison with the load, i.e. the autonomy. Recent studies have suggested that deep cycling does not limit the lifetime of battery banks having autonomy exceeding six days.[7]

In cold-climate PV systems, battery temperatures are typically quite low and autonomy is usually much greater than six days, suggesting that the batteries should have extremely long lifetimes. Field experience suggests otherwise. In recognition of this, some researchers have suggested that grid corrosion should be treated as though it was constant at temperatures below roughly 20°C.[7] This is a pragmatic approach, but it does not address a fundamental question: if neither grid corrosion nor deep cycling are limiting battery lifetime in cold climates, what is? A number of causes, including stratification, sulphation and incomplete charging, can be hypothesized, but as yet there is no conclusive argument.

This has deep implications for battery lifetime modelling in cold climates: if the ageing mechanism is unknown, how can it be modelled? Nevertheless, many sizing and simulation software include ageing models, usually based on the conventional wisdom cited above. Users should be suspicious of predicted lifetimes exceeding the lifetime expected at room temperature and must be aware that lifetime depends on factors neglected by most models.

Life-cycle cost analyses require an estimate of battery lifetime in order to determine replacement costs. Battery replacement is a major expenditure over the life of a PV system; the accuracy of this analysis is no better than the accuracy of the lifetime estimate.

Battery temperature modelling

Given that temperature strongly influences battery operation and lifetime, it would seem obvious that simulation and sizing software should include a model to predict the temperature of the battery. Surprisingly few packages do; rather, battery temperature is commonly equated to the ambient air temperature, the average ambient air temperature over the month or the average minimum daily temperature. This often results in large maximum errors in battery temperature, especially in cold climates, leading to serious errors in the sizing of the battery. Moreover, it masks the differences between uninsulated enclosures, insulated enclosures and insulated enclosures with phase-change materials, making it difficult to evaluate their relative merits.

Simple models have been developed and validated.[8] The user should investigate how their software estimates battery temperature and evaluate how much error this will introduce.

PV modules

In comparison with the battery, PV modules are simple to model: their behaviour is easily characterized, they have no memory and their characteristics change little with time. As a result, PV arrays are accurately modelled by many sizing and simulation packages.

All array models explicitly include the insolation as an input, and most software packages account for the effect of temperature on the module behaviour – an important requirement for cold climates. Already this results in a reasonably accurate model. Further refinements include:

- *Module–battery interaction.* The power output of a module depends on the voltage at which it is operating; in most stand-alone systems the battery determines this. Thus, accurate modelling of the module may necessitate accurate modelling of the battery voltage, especially when the module temperature is high or the module has a low maximum power-point voltage (e.g. 'self-regulating' modules with fewer cells in series). Energy-transfer models for the battery will be unable to do this.

- *Incidence angles.* When sunlight strikes the surface of the PV array at an angle greater than approximately 60°, reflection at the surface of the array reduces the amount of light actually reaching the cells. High incidence angles can be common for high-latitude installations and for vertical arrays (e.g. facades) at lower latitudes. Many software packages take this effect, which can range from 5 to 40% of the daily output, into account.
- *Low-light-level performance.* Many PV modules perform poorly at low light levels; below 200 W/m², module efficiency declines rapidly.[9] Radiation at levels below 200 W/m² can account for 5 to 15% of the annual energy incident on an array at mid to high latitudes and an even larger portion of the wintertime radiation.[10] Moreover, not all modules perform equally well at low light levels, even when their efficiency at standard test conditions is comparable. Few software packages currently address this.
- *Spectral effects.* When the spectrum of sunlight differs from that of standard test conditions, the efficiency of PV modules – notably amorphous silicon – changes as well. In cold climates, the spectrum will be altered by varying air mass, atmospheric composition and ground reflection. Few programs take this relatively minor effect into account.

Real-world phenomena such as snow and ice accumulation on the modules are next-to-impossible to model, but do affect the accuracy of the simulation.

Radiation modelling

All PV module models require, as input, the amount of solar radiation striking the array. Databases of solar radiation data usually contain only the total radiation incident on a horizontal surface. If the array is not horizontal, the global radiation on the horizontal must be transformed into the total radiation incident on the plane of the array.

As illustrated in Figure 11.1, there are several steps involved. First, the global radiation on the horizontal must be split into its beam and sky-diffuse components, which are treated separately in subsequent calculations. This split is achieved by models that correlate the clearness index (i.e. how much radiation is striking a horizontal surface versus how much would strike a horizontal surface outside the earth's atmosphere) to the fraction of the global radiation that is diffuse. Second, the beam radiation on the horizontal is transformed into beam radiation in the plane of the array by geometrical considerations and the diffuse radiation on the horizontal is used to estimate the diffuse radiation in the plane of the array by semi-empirical 'sky' models. Third, the albedo and geometry are used to calculate the ground-reflected radiation in the plane of the array. Fourth, the beam, sky-diffuse and ground-reflected parts of the radiation are summed to give the total radiation in the plane of the array.

Existing correlation models and sky models appear to be less accurate at higher latitudes, at least for hourly data.[11] This is not surprising, since they are derived from observations at much lower latitudes. At latitudes above 60°N, the sky models do the following:

- They tend to overestimate the diffuse radiation on the plane of the array, leading to average errors in the total radiation in the plane of the array of 5 to 30%.
- They are less accurate when the array faces away from due south.
- They are less precise (i.e. have a larger root mean square error) than at moderate latitudes.

Brunger and Hollands[11] also found that, regardless of the sky model used, one common correlation model (the 'Perez' model) led to average errors for the plane of array total radiation of 7 to 9% in a study for Alert, Northwest Territories (82°N). Since they are based on similar principles, the models dealing with monthly radiation data may echo these inaccuracies.

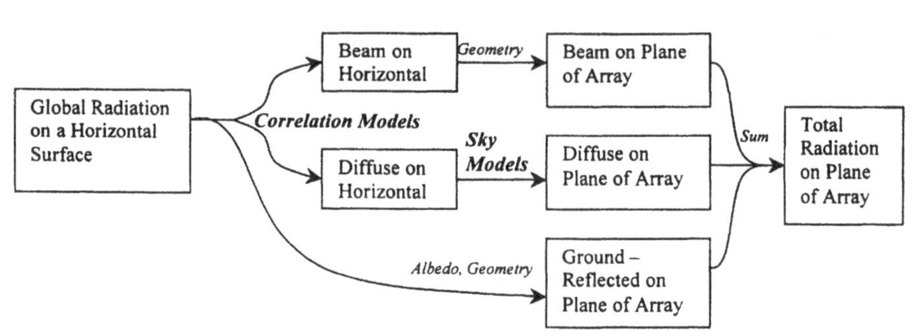

Figure 11.1. Radiation modelling. Correlation and sky models are less accurate at high latitudes

When a simulation requires hourly data but only monthly averages are available, *synthetic weather data generators* can produce artificial hourly data that mimic monitored data in terms of frequency distribution and serial correlation. Such routines usually require the average value and standard deviation for the daily global radiation on a horizontal surface.

Some software includes models for the radiation on tracking arrays, radiation augmentation by reflectors and shading by nearby obstacles. The latter may be especially useful at mountainous sites, where sunrise and sunset will often occur early because of adjacent mountains blocking the horizon. When adjacent mountains do not block the horizon, high-altitude sites will experience sunrise earlier and sunset later than lower-altitude sites, but the position of the sun in the sky is not affected by altitude.

Most simulation and sizing software works with typical or average weather data. This is useful for determining the average output for a grid-tied system, but is not ideal for sizing stand-alone systems with critical loads: here 'worst-case' radiation data for the darkest winter months are more useful. Some sizing software and many sizing software users address this problem simply by oversizing the system by some arbitrary safety margin, thereby inflating costs and defeating the purpose of using computerized tools. A better approach is to reduce monthly radiation averages by half a standard deviation;[12] synthetic weather data generators can be used to turn these diminished averages into hourly data, where needed. This is not perfect, however, because the choice of half a standard deviation is somewhat arbitrary. The best approach is to sift through 15 or more years of monitored monthly or hourly weather data and extract the data from the least-sunny month or year. Obviously, this is time-consuming and will not be possible if historical data is not available.

Temperature modelling

The ambient air temperature is used in models for array and battery temperature. Monitored temperature data is often available from the same sources as solar radiation data. Usually the monitored data comes from a site at some distance from the PV installation and temperatures at the two sites may differ. For mountainous areas, note that for every 1000 m of increase in elevation, the air temperature will drop about 6°C, but valleys and ravines collect cold air and may be much colder than surrounding areas or sites at higher elevation ('temperature inversion'). If the weather station is in a valley but the PV installation is on a hill, the PV installation may be warmer.

Monthly averages for temperature contain no information about temperature variation or extreme values during the month. This is critical to battery modelling, since extremely low temperatures greatly reduce battery capacity. Synthetic weather generators producing hourly temperature data are less common than generators for radiation data. Complications arise since temperature and radiation are correlated, and this correlation will be important to PV system operation. Subtracting one standard deviation from monthly average temperatures yields a reasonable, if somewhat arbitrary, estimate of the worst-case (coldest) winter,[12] but ignores the correlation between radiation and temperature.

Hybrid system modelling

Hybrid systems are much more complex than PV stand-alone systems and intuition is even less reliable. Moreover, hybrid systems are typically larger and more costly than PV stand-alone systems, so mistakes in sizing are correspondingly more costly, especially at remote sites. Simulation and sizing software, though not perfect, is recommended.

Considerations for modelling and simulation of hybrid systems include:

- Models should account for the higher density of cold air, which will raise wind turbine output.
- Models should account for decreasing air density at higher altitudes, which affects wind turbines and combustion engines.
- Since the power generated by a wind turbine is approximately related to the cube of the wind speed (in theory), accurate wind-speed data is crucial to accurate simulation. Often accurate wind-speed data can be obtained only by installing an anemometer at the proposed site and hub height. Hourly data should be collected for the period of a year or more.
- Rime and glaze icing of wind-turbine blades can be a serious problem for turbines sited on ridges or mountains near open water. By its effect on the blade aerodynamics, even a small quantity of ice may reduce turbine output significantly. Modelling the frequency and severity of icing is difficult.
- Cold gensets are reluctant to start and starting them causes wear; simulations may not recognize the implications of this for dispatch and control strategy, but the user should.

REFERENCES

1. Duffie JA, Beckman WA (1991). *Solar Engineering of Thermal Processes*. New York: John Wiley & Sons.
2. Gordon JM (1987). Optimal sizing of stand-alone photovoltaic solar power systems. *Solar Cells*, 20, pp.295–313.
3. Ratajczak AF (1985). Photovoltaic-powered vaccine refrigeration–freezer systems field test results, NASA Rep. TM-86972. Lewis Research Center, Cleveland, Ohio.

4. Kaiser R, Reise C (1996). *PV System Simulation Programs.* Freiburg: Fraunhofer Institute for Solar Energy Systems ISE.
5. Hill M, McCarthy S (1992). *PV Battery Handbook.* Cork, Ireland: Hyperion Energy Systems Ltd.
6. WATSUN Simulation Laboratory (1997). *WATSUN-PV 6.0 User's Manual and Program Documentation.* Waterloo, Ontario: WATSUN Simulation Laboratory, University of Waterloo, 1997.
7. Spiers DJ, Rasinkoski AA (1996). Limits to battery lifetime in photovoltaic applications. *Solar Energy,* 58(4–6), pp.147–154.
8. Ross M (1998). Estimating wintertime battery temperature in stand-alone photovoltaic systems with insulated battery enclosures. *Renewable Energy Technologies in Cold Climates '98, 4–6 May 1998, Montréal, Québec,* pp. 344–350. Solar Energy Society of Canada, Inc., Ottawa, Ontario, Canada.
9. Bucher K, Kleiss G, Bätzner D (1997). RRC Module Energy Rating: A Module Survey. *Proceedings of the 26th IEEE PV Specialists Conference, Anaheim, California, 29 Sept.-3 October 1997.* Institute of Electrical and Electronics Engineers, Inc., Picataway, NJ, USA.
10. TamizhMani G, Dignard-Bailey L, Thevenard D, Howell RG (1998). Influence of low-light module performance on the energy production of Canadian grid-connected PV Systems. *Renewable Energy Technologies in Cold Climates '98, 4–6 May 1998, Montréal, Québec,* pp.279–284. Solar Energy Society of Canada, Inc., Ottawa, Ontario, Canada.
11. Brunger AP, Hollands KGT (1995). *Solar Irradiance Modeling for High Latitudes.* Waterloo, Ontario: Solar Thermal Engineering Centre, University of Waterloo.
12. Leblanc F (unpublished). Worst-Case Winter(Temperature) and Solar Radiation Year. Natural Resources Canada, Varennes, Québec, Canada.

IV Installation and operation

12 Installation

Chuck Price
(Sovran Energy Inc.)

At some point a photovoltaic system moves off the drawing board and onto the site. Assembling the components into a *functioning* system in an efficient manner demands careful planning. In cold climates, installation is often complicated by remoteness and difficult working conditions.

SAFETY CONSIDERATIONS

Photovoltaic systems are not particularly hazardous, but like any power system, they do pose certain dangers. Before any transportation, installation, operation or maintenance activities, the installer should be acquainted with the hazards of photovoltaic systems:

- *A PV array can cause electrocution.* Any array capable of voltages exceeding 50 V – for example, an array of three or more typical modules in series or even a single AC module – should be considered dangerous. Moreover, note that a PV module cannot be turned off: whenever light reaches the cells, it will be capable of producing current. Avoid servicing the system in wet weather if possible.
- *Batteries can electrocute or cause burns.* When sufficient cells are connected in series, the batteries in a PV system can have voltages exceeding 50 V across their terminals and are thus capable of electrocution. In addition, regardless of battery voltage, batteries are capable of providing extremely large currents that will cause sparks when the battery is connected or short-circuited. These sparks can ignite flammable materials and could cause burns. Avoid servicing the batteries in wet weather if possible.
- *Batteries contain caustic liquids and heavy metals.* Battery electrolyte can burn skin, damage eyes, attack tooth enamel (leading to tooth loss), damage lungs if strong vapours are inhaled or corrode internal organs if ingested. Wear protective glasses and rubber gloves when handling batteries or adding water to them. Avoid contact between the battery and skin or clothes. If electrolyte comes in contact with skin, run water over the skin for 10 minutes. If electrolyte comes in contact with eyes, immediately flush the eyes with water for ten minutes while holding the eyelids apart, then contact a doctor. Clean up electrolyte spills by:

    1. neutralizing the electrolyte (use baking soda for lead–acid batteries and vinegar for Ni–Cad batteries);
    2. cautiously washing the area with soap and water.

    Note that battery electrolyte can react with other chemicals; keep other chemicals away from batteries.

- *Batteries evolve hydrogen and oxygen, which is an explosive mixture.* Keep flames, lit cigarettes and sources of sparks away from batteries. Batteries should be well ventilated, especially when people are working on them.

PREPARING FOR INSTALLATION

Organization for remote installations

When considering the installation of a PV system in a cold climate, the remoteness of the location must be taken into account. Since PV installations are not usually used where grid electricity is already available, most installations are in remote locations. Remote can mean at the top of a mountain, 400 km from the nearest town or 2 km across a river or lake. The installation must be carefully planned in advance. The logistics of shipping, sufficient tools, spare-part inventories, redundant components, proper clothing, temporary heat and shelter, housing, food, transportation and communication are all important considerations prior to actual installation or travel to the site.

Even if the installation is close to a small town, it is possible that miscellaneous supplies, such as wire connectors, screws, bolts, fuses, electrical tape, flashlight batteries, silicone caulking, etc., will not be available. One missing piece or tool may require an extra trip, which could double the construction time for a small project. *All* of these components must be shipped with the system.

Many locations are accessible by a cellular phone or radio, but some locations still have no communications. This makes it very difficult to order spare or missing parts or to request transportation. System pre-assembly and meticulous planning will reduce the requirements for additional components, tools or spare parts.

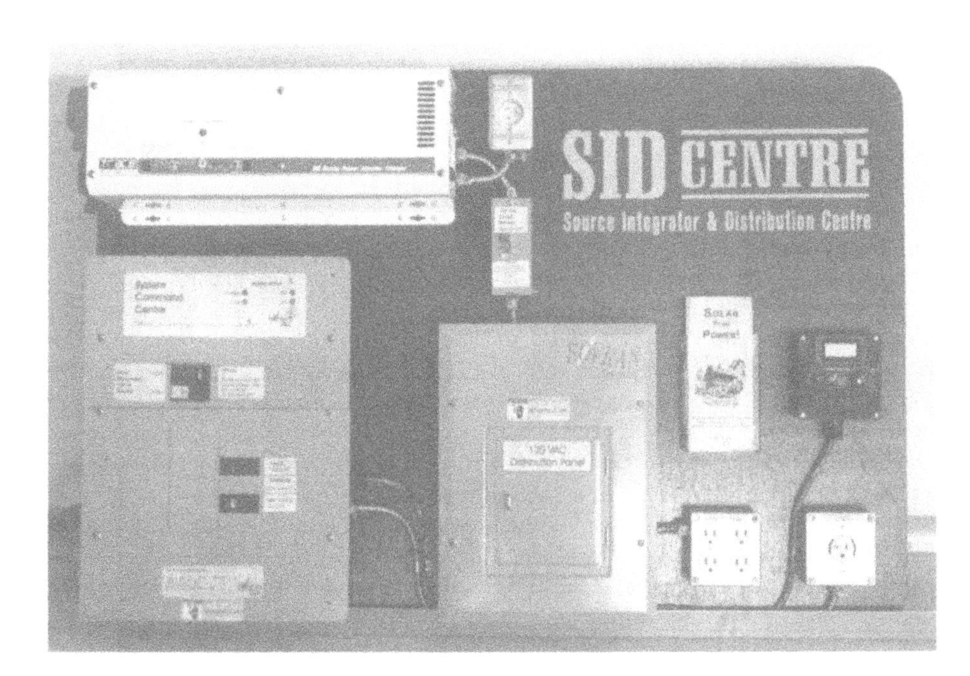

Figure 12.1. Pre-assembled controller, inverter, disconnects and distribution centre

If remoteness or climatic conditions make it difficult to access a site, it is advisable to build redundancy into the system and/or leave an extensive cache of spare parts at the site. Critical loads (loads that cannot be allowed to shut down) will justify designing redundancy into the system. If a system requires maintenance, a technician must be flown in first to determine the problem and then out to get the part and then back again to install the component. It may be much less expensive to include redundant components and copious spare parts at the site, so that fewer visits are required.

Pre-assembly

As a result of many years of experience, many project components are now pre-assembled and tested on the supplier's premises before being shipped to the site (Figure 12.1). For any PV system it is recommended that the modules and structure be pre-assembled at the vendor's shop and then broken down into skids or pallets of a size appropriate to the type of transportation to be used. The balance-of-system components, such as inverters and control equipment, should be premounted and prewired on a plywood backboard and shipped to the site as a unit. Prior to shipping, the system can be operationally tested, to avoid surprises at the site. These component blocks are then shipped to the site along with *re-assembly instructions and component literature*. Once the equipment is on site the system is reassembled and tested.

This approach has three advantages. First, pre-assembly drastically reduces the site labour. Second, it reduces the chances of small parts being forgotten or overlooked. Third, the system is tested prior to shipping and hence start-up problems are reduced. Experience has shown that for very large projects overall labour and start-up costs can be reduced by up to 50% compared to total on-site construction.

Shipment

When planning shipment, there are three key considerations. The first is to ensure that all necessary parts and equipment are shipped. Construction in remote areas can become very expensive when required components are not available. It is possible for a large project to be held up for one week because of a missing $2 item. In some locations, equipment can only be brought in once per week and then only by helicopter. In Canada, helicopter time costs between 250 USD/hr and 700 USD/hr depending on the size of helicopter. A missing piece may mean a special helicopter trip and a construction crew with zero production. Obtaining parts in remote communities, even those with a daily air service, may be difficult: in Iqaluit, Northwest Territories, for example, shipped goods are prioritized and electrical supplies may have to wait until space is available.[1]

The second key consideration is to ensure that there is no package in the shipment that is too large or too heavy for the modes of transportation that will be used. For example, ensure that if they are to be manoeuvered by humans, batteries are not too heavy to be lifted by one or two people.

The third key consideration is selecting the best time to transport equipment; this may not coincide with the best time to do installation. In muskeg areas, for example, summer transportation is by air only, whereas winter transportation may be by truck over ice roads or by snowmobile. Since winter construction can be very difficult, one possible scheme is to ship the equipment to the site in the winter and then complete the construction during spring, summer or fall. Note that, if equipment is transported in advance of installation, it must be stored in such a way that it will not be damaged by the elements. Protect plastic from the sun's ultraviolet rays and cover any materials contained in boxes or bags that are not weatherproof, such as cement in paper bags. Batteries should be shipped to the site ahead of construction only if they can be stored in a warm area or if they will stay nearly fully charged throughout the winter. Lead–acid batteries must not be permitted to freeze.

INSTALLING THE SYSTEM

Transportation and site conditions

The installer should investigate the conditions he or she can expect to encounter at the site and during transportation. By properly preparing for these conditions and, whenever possible, planning transportation and installation to coincide with the best, or least-bad, conditions, the expense and difficulty of installation can be minimized.

Weather

Weather affects people's ability to carry out a job, and cold-climate areas have weather extremes. Cold temperatures make even simple tasks, such as wiring, difficult. In cold climates, low temperatures may not be limited to winter. Moreover, temperatures do not need to be below freezing to impede installation – combined with rain or wind, temperatures above freezing make work difficult and even dangerous. Wind chill, the phenomenon whereby heat loss due to wind makes the temperature feel colder than it actually is, may be a consideration year-round. For example, to exposed flesh 0°C with a wind velocity of 25 km/h feels identical to –32°C

The temperature drops roughly 6°C with every 1000 m increase in elevation, so alpine sites may be quite cold, even when the valleys below them are warm. For example, the weather conditions at the top of Mount Assiniboine in British Columbia, Canada (51°N, 3618 m) are similar to those of Inuvik, NWT, Canada (68°N, 68 m).[2]

During summer, very hot temperatures may be problematic at some sites, especially when daylight hours are long and there is little respite from the sun.

Adequate clothing and protection from the weather is necessary. Rain gear, hiking boots, mitts, warm hats, coveralls, heavy winter jackets and blue sunglasses for snow reflection may be required. Very cold temperatures mean working with heavy gloves or mitts and very bulky clothing, slowing installation down. Some construction sites use *hoarding*, i.e. canvas or plastic tarpaulins erected temporarily as a windbreak, but this normally reduces only the wind chill factor – effective heating is extremely difficult and expensive. Labour expenses escalate dramatically for remote cold-weather installations.

Snow, cloud and fog

Falling snow, cloud and fog may interrupt helicopter and aeroplane flights and reduce visibility. At some sites, blizzards and white-outs may strike with little warning and pose real danger. Drifting snow can hide equipment and make land transport difficult. A compass or global positioning system (GPS) should be carried if this is a concern. In winter, cloud and fog may cause rime ice to form on equipment. These problems are most severe in mountainous areas.

Daylight

Just as long days during the summer may permit longer working hours, short or non-existent days during the winter may slow work down. Construction lighting may be necessary.

Wind

Extreme winds affect not only humans, but may damage incomplete installations or unanchored equipment, and make work difficult because things will have a tendency to blow away. Winds tend to be especially strong along ridges and at mountain peaks.

Insects

While summer conditions in most cases favour installation, they are not without difficulties: except at alpine sites, black flies, mosquitoes, ticks or other biting insects are usually present and mosquito repellent is not particularly effective. A better solution is to keep arms, legs and head covered with light-coloured clothing, but this can be very uncomfortable when the temperature is close to 30°C. Early spring and late fall may be preferable: the weather may be pleasant and predictable and it may be prior to or after the insect season. It is wise to consult people familiar with the site.

Animals

In some areas, curious or aggressive wild animals such as bear, moose and elk, are a concern. All animals are very protective when they are caring for their young; hence care must be taken if these animals are in the vicinity. For areas inhabited by bears, 'bear scare' is available.[3] This is a small device that, when activated, makes a loud noise that is intended to scare bears away.

Ground conditions
During winter, the ground will be frozen, making digging, burying and driving stakes difficult. During summer, mud and marshy conditions may complicate transportation and installation.

Flooding
During melt-off and during rain squalls low-lying areas and normally dry stream beds may flood; materials and equipment that are stored in such areas may be lost or damaged.

Situating the photovoltaic system

The photovoltaic system must not be shaded by long shadows cast by trees or other obstacles. A chainsaw may be required to remove trees so as to give clear access to the sun. If the site is in a park or other protected area, special permission may have to be obtained prior to removing trees.

If PV modules are visible and accessible, someone may decide that a module is more useful for charging a car battery than for its current use, especially at sites where there are few people around to observe a theft. Tamper-proof bolts and nuts and chain-link fencing with padlocked gates will deter thieves. Whenever possible, the PV array should be located so that it is not visible from adjacent roads. Many PV systems are located in remote, designated hunting areas and hunting involves target practice particularly if there is a lack of animals to shoot.

During winter, ice and snow may accumulate on guy wires, structures and arrays. Typically the ice will fall off in chunks owing to melting or wind and snow will slide off the array as a sheet. Falling snow and ice can be dangerous to maintenance and construction personnel (or, at more accessible sites, passers-by) and may damage PV arrays or other equipment. The PV array should not be situated under guy wires or above where someone is likely to walk in the winter. Guy wires can also project a narrow shadow that will significantly reduce a module's output.

Orienting the array

Exact orientation of the array can be difficult and confusing. At high latitudes the difference between true north and magnetic north can be significant – for example, at Whitehorse, Yukon Territory, Canada, there is 28° difference between the two. A compass can be used if the magnetic declination is known or is small, except when so close to the magnetic pole that it is unreliable. In this case, a global positioning device may be necessary. Alternatively, a simple method for determining the true north–south orientation is to examine shadows at solar noon (roughly noon standard time or one o'clock daylight savings time): they will point directly to true north (northern hemisphere) or true south (southern hemisphere). Solar noon can be determined by calculating the difference between the local meridian and the meridian by which time is set in that time zone or by examining the length of the shadow of a vertical object. At latitudes greater than 24°, the shadow will be at its shortest at solar noon and will run perfectly north–south.

Foundations

Foundations for mounting structures present special challenges. The terrain found at cold-climate sites may make it difficult to set a solid foundation. Even standard soil conditions can be a challenge: digging may have to be undertaken by hand because excavation equipment may not be available; concrete may be impractical because of freezing conditions, lack of water or exorbitant shipping costs. Prior to designing the support structure, review the site ground conditions and the resources available at the site in terms of the availability of steel, wooden or concrete pilings and/or the possibility of using local timber for sleepers (beams placed horizontally in or on the ground as a foundation) (see Table 12.1).

For many muskeg areas, wooden or steel pilings can be used; in some locations pilings cannot be used as wooden

Table 12.1. Foundations for various types of terrain

Type of terrain	Foundation	Comments
Rock	Poured concrete or sleepers	Drill holes into rock to anchor sleepers.
Permafrost	Pilings if drilling equipment is available; otherwise anchored sleepers or platform frozen into permafrost	Permafrost areas generally have a non-frozen surface during the summer. The top 10 cm are very environmentally sensitive. *Disrupt as little as possible.*
Muskeg/swamp	Wooden or steel pilings or a log/timber floating platform	Wooden pilings may float; a solid base may never be reached. For platforms, pay attention to the centre of gravity and anchor with guy wires.
Sand, clay, black soil	Concrete, wooden or steel pilings or sleepers; poured concrete	For soils that contains no rocks, structures supported by a platform can be anchored using guy wires attached to dead man anchors screwed into the soil (Figure 12.2). (A dead man anchor is installed in the ground at the end or corner of a transmission line. For soil containing little rock, a plate type of dead man can be screwed into the soil by hand or with a lever.)

Figure 12.2 Installation of a plate dead man

pilings may float or a solid base is not reached regardless of how deep the pilings are driven. An anchored, floating platform is a possibility for very wet areas.

In cold-climate areas where permafrost does not occur, soil can freeze to a depth of 2 m in the winter. Foundations must extend below this frost level or the foundation will shift over time due to the expansion and contraction of the soil as it freezes and thaws.

Pilings and sleepers can be used in permafrost areas; another foundation that has been employed successfully for photovoltaic systems involves freezing a platform directly into the permafrost.[4] During summer, the site of the foundation is excavated to the permafrost level. After exposure to the ambient air, the top 30 cm of permafrost thaws and can be removed. A wooden or steel platform is then placed in the permafrost, the structure is bolted to this platform, the top of the platform is insulated with 10 cm of foam insulation and then the fill is replaced. The platform will freeze into the permafrost and be immobilized.

Electrical considerations

Most electrical issues are dealt with at the design stage. On the other hand, details related to electrical cables and grounding may be left to the installer. At very low temperatures wire insulation becomes brittle and can be damaged easily. The wire used should have a cold temperature rating and be resistant to the ultraviolet rays contained in sunshine, especially at high altitudes. Wholesalers familiar with the area of the installation should furnish appropriate wire. Installation should not be undertaken at temperatures lower than –20°C.

Because the terrain in cold climates can be permafrost, rock or muskeg, grounding can also be a major issue. For permafrost or rocky terrain, use grounding plates or attach to steel pilings if they exist. In muskeg, satisfactory grounding can be achieved using ground rods. The entire system must be grounded at only one point.

Protecting equipment from insects and animals

The installer must ensure that insects and animals will not damage the system after he or she leaves. Most animals are inquisitive and many like to scratch their backs and sides on metal edges. Structural supports are attractive to itchy elk, moose and bear. Loose wires can be a plaything for bears. Polar bears are particularly inquisitive. Ground conditions permitting, a chain-link fence with a locked gate may be a good investment. Smaller animals such as mice and weasels may try to make their home in battery or control-equipment enclosures; ensure that the design and construction of these enclosures will deter these guests. Some animals also gnaw on plastic wire insulation. Making cables inaccessible or using armoured conduit will avoid this problem.

Insects will penetrate and fill up all openings. Electrical boxes with open holes or vents will eventually get plugged with insects. Ventilation holes and the holes in electrical boxes and conduit should be covered by a screen.

REFERENCES

1. KRT Electric Staff, Iqaluit, Baffin Island (1997). Personal discussions of electrical installations in the Arctic, August.
2. Atmospheric Environment Service (1997). *The Climate of Contiguous Mountain Parks, Banff-Jasper -Kootenay-Yoho Report*, #30. Toronto: Environment Canada.
3. Price J (1997). Aerial Recon Surveys: Personal discussions of operating conditions on a gas gathering site in Northern Alberta, July.
4. Egles D, Wissink R, Usher E (1995). PV power in the high arctic. *Proceedings of the 21st Annual Conference of the Solar Energy Society of Canada, Inc., Toronto, Ontario, 31 October–2 November.* Solar Energy Society of Canada, Inc., Ottawa, Ontario, Canada.

13 Operation and maintenance

Dave Egles
(Soltek Solar Energy Ltd)

This chapter examines the operation and maintenance of a photovoltaic power system after it has been installed. Although most PV systems operate automatically and without the need for user intervention, operators must still consider what is required to service and maintain the system over the years. Some adjustments may be made seasonally, but most servicing is performed as preventative maintenance. This chapter looks at the operation of solar systems, typical maintenance procedures, the diagnosis of problems and equipment replacement intervals. Note that PV systems in cold climates are operated and maintained in largely the same way as PV systems in warm climates.

Before performing any operating or maintenance procedures, one should review the summary of safety warnings and considerations in Chapter 12 and read equipment manufacturers' warnings.

PV POWER SYSTEM OPERATION

One of the characteristics of most PV power systems is that they operate automatically, i.e. without user intervention. If properly designed, users may not visit a site for years. Battery charging, regulation and load control are managed according to parameters programmed into the control equipment. These set points are rarely changed after system commissioning and a properly sized power system will operate the loads without failure. However, sometimes seasonal changes are made to the power system, as discussed below.

Array tilt angle adjustments

Some users change the tilt angle of the solar array to maximize the output by season. As discussed in Chapter 3, this makes sense only when it is critical to maximize electrical energy production *at all times of the year*, as may be the case with cottages or remote residences. For example, at a latitude of 60°, the ideal panel mounting angle is 30° from the horizontal in the summer (May through August) and 75° in the winter (November through February). Adjustment of the tilt angle twice a year (to 30° in April and 75° in September) would result in roughly an 18% increase in power output over an array left at the winter tilt angle of 75°. On the other hand, with grid-tied and hybrid systems, it is the *total energy production over the year* that should be maximized and the additional energy produced by adjusting the array twice rather than leaving it fixed at the optimal tilt angle for year-round operation (e.g. 45° at a latitude of 60°) is only 3 to 5%. If the power system is designed to operate *a specific fixed load year round* (as in most stand-alone PV systems), then the tilt angle is fixed to maximize winter output and adjusting the tilt angle makes no sense, since there will already be surplus production during the summer. Note that adjusting the array four or more times a year has nearly negligible additional benefit compared to biannual array adjustment.

For those systems which track the sun, it is often wise to turn off the tracker and fix the array in place for the winter, especially at higher latitudes. The additional energy that a tracker will make available during the winter is minimal, since the sun's position changes little in the sky. Moreover, cold temperatures may reduce tracker reliability (see Chapter 5).

Operation of PV systems with primary batteries

Some PV systems may have a caustic potash primary (non-rechargeable) battery bank to back-up the solar system. This type of system typically uses a relay to switch the load to primary battery power when the battery voltage drops below a certain threshold. The potash batteries thus prevent discharge of the lead–acid batteries below the point at which their electrolyte is vulnerable to freezing. For example, at a certain site –20°C battery temperatures are anticipated. At this temperature, the lead–acid batteries may freeze if discharged below 50% of their nominal capacity (see Chapter 6). In this system, the potash batteries would be set to switch in at a voltage corresponding to 40% depth of discharge to protect the lead–acid batteries.

Hybrid power system operation

PV–genset hybrid systems have their own operational issues. Starting the genset at cold temperatures can be difficult and causes wear to the genset. In order to avoid this, careful design of the system must be accompanied by attention to the strategy used to control when the genset is started. The genset is usually programmed to start automatically based on three conditions:

- when the battery state of charge drops to a certain threshold, typically 50% or higher;
- when the temperature of the genset room drops to a certain temperature, usually no lower than 0°C;
- at a fixed interval to exercise the generator, typically every two to four weeks.

PV–genset systems may also have some degree of user interaction in the form of remote generator starting. While usually automatic, some systems feature manual override of programmed settings and on-demand starting of gensets. At remotes sites these can be activated through a radio link.

Removing snow and ice accumulation

Other operational procedures that may be necessary with some cold-weather PV systems include removal of snow and ice from the arrays. Snow generally melts off arrays fairly rapidly, but icing in mountainous and Arctic regions may be more tenacious and demands attention during system design and operation (see Chapter 5). In most cases, stand-alone PV systems will be designed with sufficient battery storage to operate loads safely through the winter, even if there are extended periods with ice and snow covering the solar panels. As an extreme example, some Canadian mountain-top sites accessible only by helicopter have 120 days of battery autonomy in order to guarantee that there is power to the load. On the other hand, some users may be willing to accept the occasional winter service visit to remove ice from solar panels in order to avoid needing a large, costly battery bank.

Preparing summer-only PV systems for winter

Systems that operate only during the summer, such as those at cottages, should be prepared for winter to ensure that batteries and arrays are not damaged during the winter. Batteries may either be disconnected from the system or left connected to the system with all loads turned off. In either case, the battery must remain nearly fully charged throughout the winter to prevent damage due to the electrolyte freezing. If the batteries are removed from the system, ensure that they are fully charged when stored and that self-discharge prior to winter does not significantly reduce the state of charge; during winter, when the battery is cold, self-discharge will proceed extremely slowly and is not a concern. If the battery is left connected to the system, ensure that no loads are left on, including minor 'hidden' loads such as the stand-by power consumption of an inverter. Check battery electrolyte levels at the end of summer and the end of winter; without a load, the array will keep the battery near its gassing voltage throughout the winter, possibly leading to water loss.

If the array tilt angle is very low, ensure that winter snow loads will not cause failure of the modules or structure. If the tilt angle is adjustable, it may be increased to improve the array's ability to shed snow.

MAINTENANCE

Maintenance frequency

Maintenance of solar power systems is minimal and in many cases users will not visit sites for years at a time. Frequency of site visits is usually determined by the service interval of the load equipment or the batteries, with annual site visits being the norm. Typical maintenance intervals for batteries commonly used in PV systems ranges from semi-annually to every two or three years. Sites with flooded batteries will require a site visit at least once a year to ensure that any plate material has not been exposed as a result of electrolyte evaporation, as exposed plate material will oxidize and permanently lower the battery capacity; in warmer climates more frequent visits will be required. With maintenance-free batteries, like gel cells or absorbed glass mat batteries, an annual service visit is recommended but not mandatory. If batteries dictate the frequency of site visits, and site access is difficult, then the case for installing maintenance-free batteries is a good one.

Typical maintenance procedures

If an annual site visit is scheduled, there are a number of activities that can be performed to ensure the proper long-term operation of the power system. While most tests and checks are at the discretion of the users and may be necessary only if a problem is suspected, certain procedures are essential, namely those associated with battery maintenance. Typical routine maintenance procedures for PV power systems are described below and summarized in Table 13.1.

Solar array

- *Clean solar modules*. Remove bird droppings, leaves and dirt to eliminate shading of individual solar cells, which can significantly reduce module output and create damaging 'hot spots'.
- *Inspect all wiring and connections.* Thermal expansion and contraction can loosen connections, so these should be retorqued to manufacturer's specifications, particu-

Table 13.1. Typical maintenance activities for PV power systems

Component	Maintenance action	Frequency	Priority*
Solar arrays			
Solar panels	Inspect and clean	Annually or with site visits	3
Wiring	Inspect wiring for abrasion, secure attachment	Annually	3
Terminals	Torque as per manufacturer's specifications	After first winter and as required	1
Array output	Open-circuit voltage and short-circuit current test	Annually or with site visits	2
	Compare output of each subarray		
Controllers			
Set points	Confirm voltage set points	Bench test if problems suspected	3
Relays	Verify charge relays are closed and current is delivered to batteries. If regulating, reset and observe regulator function	Annually or with site visits	2
Batteries			
All types	Record cell voltages, impedance	Annually	2
	Terminal torque	Annually	1
	Capacity test	If problems suspected	3
Flooded lead–acid	Inspect electrolyte levels	Annually or more frequently	1
Nickel–cadmium	Replace electrolyte	Per manufacturer's recommendations	3

* 1 = essential, 2 = highly recommended and 3 = recommended,

larly after the first winter. Look for loose wires, wire strands that bridge two connections and signs of heating such as discolouration. If loose connections are found, be careful: there may be dangerous voltages between the two points where the connection is broken as well as between either point and ground.

- *Array voltage.* Measure array open-circuit output voltage using a DC voltmeter, i.e. the voltage across the array terminals when the array has been disconnected from the rest of the system. Exercise additional caution whenever three or more PV modules are connected in series – this leads to a voltage exceeding 50 V, which can cause injury or death. Open-circuit voltage also depends on cell temperature; it should be in the range of 20 to 25 V for a 12 V system whenever there is appreciable light. This test will indicate whether the modules are connected, but will not indicate the soundness of connections, nor whether a single panel has a problem such as a blown diode.
- *Measure the array current.* Since this will vary with the intensity of the sunlight, it is easiest to do this on a bright, cloudless day with the array squarely facing the sun – under these conditions, the intensity of the sunlight on the array should be about 1000 W/m^2. For small arrays, i.e. arrays with current less than 10 A and system voltage less than 50 V DC, the short-circuit current of the array can be measured by disconnecting the array and then *carefully* short-circuiting the array with an appropriately fused in-line DC ammeter. Do not attempt this with larger arrays or with the array still connected to the rest of the system: this could cause injury and/or equipment damage. Do not disconnect larger arrays; rather, measure the operating voltage across the array using a DC voltmeter and the current under load using a clamp-on DC ammeter. By examining the *I–V* curve given in the module specifications, the current that the array should be producing can be estimated on the basis of the operating voltage (obviously, if the short-circuit current has been measured, the voltage is zero). Compare this to the measured array current. If the measured output is lower than expected, check all wiring connections and, if these are sound, measure the operating voltage and current of each module individually. Except for small arrays, use a clamp-on meter even for individual module measurements. In larger systems, comparing the output of identical subarrays may reveal wiring problems – the current should be similar in all subarrays.

Controller

- *Verify that the controller is functioning properly.* LEDs or meters on the controller should indicate that current is flowing into the batteries. Verify this with an ammeter if necessary; for small systems an in-line DC ammeter can be used, but if it is possible that the charge current will exceed 10 A or if the system voltage is higher than 50 V DC, use a clamp-on meter. If little or no current appears to be flowing to the battery despite strong sunshine, check whether the controller is regulating, i.e. the battery is full. If the controller is regulating, reset the charge circuit by disconnecting the array for several minutes using the array breaker. The battery voltage will drop off. Turn the breaker back on and observe the battery voltage and controller indicators: when charging resumes, the battery voltage should rise to the end-of-charge voltage (see regulator specifications for this value and remember to correct for temperature). If problems are suspected, check with the manufacturer beforehand to determine whether a controller failure typically disconnects the array from the battery or leaves it connected even when the battery is full. In the former case, failure will result in low battery voltage and no charge current. In the latter case, failure will lead to battery voltages higher than the end-of-charge voltage. Bring a spare controller if problems are suspected.

- *Inspect and tighten all connections.* In low-voltage systems (less than 50 V DC), feel for hot connections, indicating a loose wire.

Batteries

- *Cell measurements.* Ideally, the voltage, specific gravity (for lead–acid batteries) and even impedance of all cells would be individually measured and recorded with each site visit. These measurements help track battery aging and can be used to identify defective cells. Individual cell voltages should all be the same – deviations of greater than 0.05 V indicate a potential problem. Discrepancies of greater than 30 points in the specific gravity of different cells (i.e. 1.20 versus 1.23) may also indicate a problem. An equalization charge (purposeful overcharge) may eliminate cell differences. If differences persist or worsen, a capacity test on the suspect cells can reveal whether the battery is nearing the end of its life (see *Battery replacement and disposal* below). Capacity testing is time-consuming and demands an auxiliary power source and load; many users find it easier simply to replace a suspect cell.
- *Battery terminal torque.* With large seasonal temperature fluctuations, and the tendency of lead to flow under compression, battery terminals require regular retorquing, preferably every year. Manufacturers typically recommend 11.3 Nm (newton metres) of torque.
- *Battery terminal cleaning.* Corrosion deposits may collect on battery terminals as a result of battery gassing. If these are present, remove the cables interconnecting the cells and then clean the battery terminals and interconnects with a stiff brush; before replacing the interconnects, grease them with petroleum jelly or anticorrosion battery grease.
- *Battery case cleaning.* Deposits of dirt or oil on the battery case can create a weak short between battery terminals. Deposits should be removed with a clean rag.
- *Battery equalization.* Periodic controlled overcharging to bring all cells to an equal state of charge and to destratify the electrolyte is sometimes recommended by battery manufacturers. Some PV controllers have automatic equalization cycles, which at a fixed interval (every 21 days for example) raise the terminal charge voltage. Other controllers may have a manual equalize switch.
- *Battery electrolyte levels.* Flooded batteries should be inspected during each site visit and battery electrolyte levels topped up with distilled water if low. Do not overfill, especially with lead–acid batteries: the volume of the electrolyte in a lead–acid battery changes with state of charge, so if the battery is full of electrolyte but only partially charged, the electrolyte will overflow when the battery is fully charged. Leave 2–3 cm of free space at the top of the battery. Battery acid should only be added if lost due to a spill. Loss due to evaporation and gassing from overcharging requires only water.
- *Ni–Cd rejuvenation.* The electrolyte in Ni–Cd batteries becomes contaminated with carbon over a period of many years. Whitish deposits in the electrolyte may indicate problems. Ni–Cd battery suppliers may also sell test kits. Replacing the electrolyte can extend battery life by up to 10 years. The typical replacement intervals are every 10 years.

POWER-SYSTEM TROUBLESHOOTING

The first signs of problems with a PV power system will be unexpectedly low battery state of charge (if monitored) or failure to power the load. In the latter case, a site inspection may reveal discharged batteries or cracked battery cases due to freezing. When problems occur with a solar power system, the most likely causes are:

- loose wiring and poor connections;
- changes in the load beyond design specifications or improper sizing of array and battery;
- declining battery performance due to cycle or shelf life or cataclysmic failure (e.g. freezing).

Note that solar modules themselves are rarely a problem as the failure rate for solar modules is negligible. If the solar array performance is suspected, wiring connections are usually the fault. In some cases lightning strikes can damage a solar module, but the damage is usually obvious. Lightning-induced surges can also destroy the shading (by-pass) diodes and leave a short-circuit path through some of the solar cells. When this occurs, the panel will have an open-circuit voltage of zero or half the rated voltage and the diode will appear damaged. The diodes can be replaced to rectify the problem.

When a problem is suspected, typical diagnostic procedures include:

- *Inspect PV arrays.* Look for loose wiring, hot connections or melted wires indicating poor connections. Inspect all wiring for breaks or cuts. Measure array current and voltage (see above). Inspect and test diodes in module junction boxes.
- *Verify that the controller is functioning properly.* Check all relays and charge circuits. Inspect temperature-compensation probes for damage. Inspect circuit boards and look for burnt components.
- *Batteries.* Check electrolyte levels, inspect seals for leakage, look for plate growth (indicated by bulging battery cases). Measure and compare specific gravity, cell voltage and impedance in all cells. Clean and torque all connections.
- *Measure the load and compare to specifications.*

BATTERY REPLACEMENT AND DISPOSAL

The batteries used in PV systems have a finite life of between 3 and 20 years, while solar panels may have lifetimes in excess of 20 years. Replacing the battery bank may be part of the maintenance schedule.

Battery life is limited by either the battery's cycle life or shelf life. Even deep-cycle batteries have a limited number of charge/discharge cycles before they wear out. This is a function of the depth of discharge and varies by brand. On the other hand, batteries used in cold-weather PV power systems are usually quite large and battery life may be dictated by shelf life rather than cycle life. Exposure to high temperatures during the summer can also have a deleterious effect on battery life. Typical lifetimes for cold-climate PV systems batteries are given in Table 13.2.

The symptoms of declining battery performance include the rapid rise of voltage while charging, wide variations in specific gravity and voltages among cells within a string, bulging cases due to plate growth, and cloudy electrolyte. A capacity test of individual cells is the only sure way to evaluate the state of the battery. This involves bringing the batteries to a 100% state of charge by an equalization charge and then discharging the batteries with a fixed load until the cut-off voltage is reached (typically 1.75 V per cell for lead–acid batteries). The total amount of ampere-hours removed from the fully charged battery to the cut-off voltage is the battery capacity. If the battery tests at less than 80% of its original capacity at the same discharge rate and temperature, then the risk of battery and system failure is increased. Most industrial users therefore retire a battery if less than 80% of the original capacity is available; in contrast, residential users may find very aged batteries perform acceptably. The user must assess the impact of a site failure to determine whether pre-emptive battery replacement is warranted. Following the capacity test, either return the cells to the battery and then perform an equalization charge to ensure that all cells are at the same state-of-charge, or replace the entire battery bank. Mixing new and used cells is not generally recommended, because the old cells tend to be overdischarged, accelerating their deterioration.

Table 13.2. Typical battery lifetimes

Type	Lifetime
RV/Golf-cart lead–acid batteries	3–5 years
Sealed flooded lead–acid batteries	3–7 years
Industrial flooded lead–acid batteries	5–15 years
Sealed industrial lead–acid batteries	8–15 years
Ni–Cd batteries	10–20 years

Once replaced, the spent battery cells must be properly disposed of. Since batteries contain caustic chemicals and environmentally dangerous trace metals, *they must be removed from the site.* They are also considered hazardous materials and must be transported to the proper disposal or recycling facilities, usually by truck. Typical procedures for PV battery disposal include:

- *Lead–acid.* Lead–acid batteries are considered dangerous goods and must be transported with appropriate safety precautions. The batteries are recyclable at the proper disposal facilities, where the acid is neutralized and the lead removed.
- *Nickel–cadmium.* Ni–Cd batteries contain potassium hydroxide, which must be neutralized before recycling. Cadmium is considered an environmental contaminant and requires special treatment. Ni–Cd battery suppliers offer to dispose of their customers' batteries. Like lead–acid batteries, Ni–Cd batteries are considered dangerous goods because of the caustic nature of the electrolyte.
- *Potash batteries.* Air-depolarized potash batteries may contain mercury and a solution of potassium hydroxide, so they must be transported to proper disposal facilities in sealed barrels to prevent spillage. The potassium hydroxide is neutralized and the mercury is separated and recycled.

14 System monitoring

Heinrich Haeberlin
(Berner Fachhochschule)

Monitoring a PV system permits examination of its behaviour under real operating conditions, which may be quite different from the assumptions and simulations made during design. Monitoring clearly shows if the PV system is operating according to the design goals, if it is over- or undersized and if there are potential problems that do not yet affect the operation, but may cause problems in the future and therefore demand preventative countermeasures. Monitored data provides designers with valuable information on how to improve future PV systems. It is also useful when assessing the potential and reliability of PV technology for a given application at a certain location.

However, monitoring always means *additional complexity of the overall system, additional energy consumption* for the monitoring equipment itself and a (possibly substantial) *increase in cost of installation and data processing*. Therefore it makes no sense to monitor every PV system.

It is very useful for the designer or the owner to monitor a PV system if there is something new in it, e.g. a new configuration, new components or a new location with an unfamiliar climate. It is also a good practice to do at least some monitoring of large PV systems (many kW_p) or systems with very high reliability requirements.

To demonstrate that the design goals of a PV system have been met, it is not necessary to monitor the system forever. A period of two years of monitoring is sufficient in many cases. Continuous monitoring over many years is required only if long-term data about reliability, degradation and/or ageing of components or energy yield are needed.

GLOBAL AND ANALYTICAL MONITORING

To keep the cost of monitoring low, a general assessment of the performance of a PV system can be made using simple *global* (or coarse) *monitoring*.[1] In this case, only *energy quantities* (e.g. irradiation in the plane of the array, energy from the PV array, energy from auxiliary energy sources and energy supplied to the loads) are determined. Suitable counters and periodic manual readings (e.g. every day or at least every month) may be used. In this case, some manpower is required for global monitoring, making it feasible only at PV installations where personnel are regularly available. It is also possible to use very simple data loggers.

Analytical (or fine) *monitoring* is a more sophisticated approach, giving much deeper insights into the operation and performance of a PV system and making unattended monitoring possible. For analytical monitoring an automatic data acquisition system is used. This may be a dedicated computer with suitable peripherals and software or a data logger. For analytical monitoring it is common to sample instantaneous quantities like irradiance in the plane of the array, temperature, voltage, current and power at regular intervals, usually between 1 and 60 seconds long. These sampled values are averaged in order to limit the amount of data to be stored and processed. The averaging period is usually between one minute and one hour. As well as such average values, it may be useful to store some peak values of irradiance, voltage, current and power in order to analyse system performance under peak load or irradiance conditions. With a suitable modem it is possible to transmit the data to a remote location for further processing.

PARAMETERS TO BE RECORDED

In the document *Guidelines for the Assessment of PV Plants* a set of parameters for the monitoring of all kinds of PV systems is proposed.[1] In many cases not all parameters need be monitored; the choice of parameters depends on the complexity of the PV system and the kind of monitoring to be performed. For instance, if only the proper operation of a stand-alone system must be verified and overall efficiency is not of interest, it may be sufficient to monitor a reduced set of parameters, such as system voltage at the battery, AC voltage at inverter output (if applicable) and some DC currents. On the other hand, in many monitoring projects performed by the author, it has been very useful to record the cell temperature in order to calculate the losses or (in very cold places) gains due to the difference between the true cell temperature and the STC temperature of 25°C. In cold climates, battery temperature should be monitored as well, since temperature is critical to battery operation. At locations where no meteorological data are available from a weather station, monitoring the global irradiance in the horizontal plane will facilitate comparisons with other sites.

Table 14.1 presents a list of parameters similar to that in the Ispra *Guidelines*. The list is applicable to analytical monitoring projects; for global monitoring, counters can be used to integrate the values of the most important variables. The three right-most columns indicate variables that should be monitored in simple autonomous PV systems ('SAPVs', with battery and DC loads only), hybrid autonomous PV systems ('HAPVs', with a maximum-power-point tracker, battery, AC loads and back-up generator) and for grid-connected PV systems ('GCPVs', without battery).

In hybrid systems with a wind generator, it may be interesting to monitor wind speed by means of an anemometer. The operating time and fuel consumption of the genset in a hybrid systems may also merit monitoring.

COMPUTERS OR DATA LOGGERS?

Reliability of monitoring equipment is sometimes lower than that of the PV system to be monitored, causing data losses. Incomplete data sets obviously make it difficult to assess system performance. Therefore *sufficient reliability of the monitoring equipment and its power supply is very important* and a key consideration when assembling an automatic data acquisition system.

Computers are very useful for larger installations, where sufficient electrical power is available. However, most computers do not operate at very low temperatures and their overall reliability (including storage devices) is usually lower than that of modern data loggers. For monitoring PV systems in cold climates computers can therefore only be considered for larger installations with a thermal management system where personnel are available at least from time to time.

Data loggers are the best choice for unattended remote monitoring, especially in cold climates. They can operate in a wide range of ambient temperatures (–25°C to 50°C is a standard, –55°C to 80°C may be available on request). If a suitable combination of modem and data logger is used, data losses due to communication problems between data logger and modem will be minimal, and reliability will be higher than that of computer-based systems.

A point to be considered is the accuracy of the internal time reference. In analytical monitoring, the time at which data is recorded is usually saved along with the data, so that the operation of the system can be accurately related to the position of the sun in the sky. Using the internal clock of a computer is very simple, but not very accurate (errors of up to 100 ppm or nearly 1 hour per year are possible). Periodic adjustment of the clock may be necessary, which is troublesome in remote installations. In Central Europe it is possible to automatically adjust the clock of a computer or data logger by means of a special receiver tuned to a time transmitter (DCF-77) located in Germany and operating at a low frequency (77.5 kHz).

REMOTE MONITORING

It is very convenient to have the monitoring data automatically transmitted back to the home office. The analyst can

Table 14.1. Set of parameters to be recorded (symbols according to the Ispra Guidelines;[1] parentheses indicate optional parameters)

Parameter	Symbol	Unit	SAPVs	HAPVs	GCPVs
Global irradiance in the horizontal plane (optional)	G_h	W/m²	(•)	(•)	(•)
Total irradiance in the plane of the array	G_I	W/m²	•	•	•
Ambient temperature in the shade	T_{am}	°C	•	•	•
Cell temperature, on reference cell or at back of modules (optional)	T_C	°C	(•)	(•)	(•)
Battery temperature (if applicable)	T_B	°C	•	•	
Array output voltage (if different from V_S)	V_A	V		•	•
Total array output current (total)	I_A	A	•	•	•
Maximum power point tracker or DC/DC converter output current (if applicable)	I_C	A		•	
Current to storage battery (+ to , – from battery)	I_S	A	•	•	
DC line voltage (battery voltage)	V_S	V	•	•	
Current to all DC loads	$I_{L,DC}$	A	•		
Inverter/rectifier DC input current (+ to, – from device)	I_{II}	A		•	
Inverter/rectifier AC output power (+ from, – to device)	P_{IO}	kW		•	
Power to all AC loads	$P_{L,AC}$	kW		•	
Power to utility grid (+ to, – from grid)	P_{TU}	kW			•
Power from backup AC generator (if applicable)	$P_{BU,AC}$	kW		•	
Current from backup DC generator (e.g. wind, if applicable)	$I_{BU,DC}$	A		(•)	
Non-availability to load, i.e. hours when load could not be powered (measured or calculated)	t_{NAV}	h	•	•	•

easily see how the PV system operates and, if there are problems, appropriate countermeasures can be taken in a timely fashion. It is even possible to make remote changes in the monitoring program. However, if data are to be transmitted from a remote location with a modem, the data logger must be carefully selected for this purpose. There are data loggers that have sporadic communication problems even with the modem recommended by the manufacturer.

At locations where the PV plant is used to power a telecommunications repeater station, a service channel for such data transmission may be readily available. At other very remote locations, radio communication is possible. The radio equipment to be used (including the antenna) must also withstand the harsh environmental conditions and will increase power consumption and cost. Care must also be taken to avoid electromagnetic interference affecting the monitoring process, especially in the case of high frequency (HF) radio transmissions (about 3 MHz to 30 MHz) – cables from the sensors may act as antennas for such signals. Filtering each wire entering the monitoring cabinet may be necessary. Similarly, if the PV system contains power converters (DC/DC or DC/AC) switching at high frequencies with insufficient internal filtering, noise may be generated, reducing the sensitivity of the radio receiver and necessitating additional filtering within the PV system itself.

ENVIRONMENTAL CONSIDERATIONS AND POSITIONING OF MONITORING EQUIPMENT

Sensors for irradiance and temperature must be exposed directly to the outside weather conditions and the insulation on these sensor cables must be specified for outdoor use and the low temperatures expected at a cold location. As meteorological services use such sensors worldwide, such devices with temperature specifications down to –55°C are readily available on the market, but they are quite expensive. Sensors for temperature and irradiance in the horizontal plane are often placed on top of a pole to avoid snow accumulation, but in the mountains the danger of lightning strikes must be considered and appropriate countermeasures (e.g. lightning protection rods, shielded cables and surge arrestors) must be taken. To give optimal correlation between the output of the plane-of-array irradiance sensor and the array output, the sensor must see exactly the same irradiance as an average module in the array; it is a good practice, therefore, to place this sensor right at the PV array, mounted exactly in the array plane, and not on top of a pole.

If wind speed is measured by means of an anemometer, it is advisable that the anemometer be heated in order to get accurate results. This will increase the energy consumption of the monitoring equipment considerably.

The data acquisition system should be located in a weatherproof cabinet, elevated off the ground to avoid flooding when the snow melts in spring. At some locations a desiccant such as silica gel is necessary in order to protect against excessive humidity. The wide range of temperatures at which data loggers can operate is adequate for most applications, even in very cold climates, and no special thermal protection is required for the data logger. On the other hand, auxiliary components, such as isolation amplifiers, current sensors and modems, may have a narrower range of acceptable operating temperatures. Moreover, accuracy is reduced when temperature variations are high. Thus if some thermal management system is used to reduce temperature variations for important parts of the PV system (e.g. for batteries), there are good reasons to let the cabinet for the monitoring equipment benefit from this as well, provided that no corrosive vapours from the battery reach the monitoring equipment.

IRRADIANCE SENSORS IN COLD CLIMATES

When PV systems are monitored, irradiance is the most difficult quantity to measure accurately. There are three main types of sensors that measure total hemispheric irradiance:

- *Thermopile pyranometers.* These estimate the irradiance by comparing the ambient temperature to that of an absorptive surface exposed to the sun's rays; the absorptive surface is protected from the wind and other elements by one or two glass domes.
- *Reference cells.* These flat, glass-covered silicon PV cells range from 8 to 100 cm^2 in area.
- *Photodiodes.* These are inexpensive light-sensitive silicon devices having a very small surface area.

Thermopile pyranometers, used worldwide by meteorological services, are highly accurate but expensive. Sensitive to all light with wavelengths of 0.3μm to 3μm, they measure nearly all power coming from the sun and make it easy to calculate the true efficiency of the conversion from sunlight to PV electricity. Thermopile pyranometers calibrated at the same time by the same institution usually agree to within 1% at 1 kW/m^2, although even the best devices will differ by up to 3% at 1 kW/m^2 when calibrated by different institutions.[2] Output voltage is of the order of a few millivolts at 1 kW/m^2 and is generally highly linear (maximum deviation less than 0.5%), but shows some temperature dependency. For cold climates, temperature-compensated devices with low temperature coefficients (e.g. the Kipp & Zonen CM21) should be used. Although thermopile pyranometers are usually suitable for horizontal or plane-of-array measurements, the accuracy of some pyranometers is compromised when they

Figure 14.1. Heated and ventilated pyranometer (tilt angle 30°) free of snow after snowfall of about 25 cm

are mounted in a non-horizontal orientation; the manufacturer's recommendations should be consulted.

Pyranometers require some maintenance. Silica gel desiccant, which combats condensation inside the pyranometer dome, must be replaced periodically, with the frequency of replacement, typically once or twice a year, dependent on the humidity. Calibration should be checked annually. For measurements of the utmost accuracy, the pyranometer should be inspected and cleaned with a soft cloth every day.

During night thermopile pyranometers will often indicate negative levels of radiation of a few W/m^2. This results from heat losses to the sky, which are especially pronounced under cloudless conditions. When irradiation values are calculated by integration, it may be desirable to attempt to compensate for this offset; as a minimum, negative night-time values should be ignored. For more information about pyranometers see Imamura *et al.*,[3] McArthur *et al.*[4] and Wardle *et al.*[2]

For measuring in-plane irradiance, a calibrated silicon reference cell is a much cheaper solution. *Crystalline silicon cells are sensitive only to light with wavelengths less than 1.1 μm.* The cells are calibrated at certain reference conditions (1kW/m^2, 25°C, standard spectrum) and under these conditions should indicate the same irradiance as a pyranometer. The spectral composition of natural sunlight varies, however, and the artificial light sources sometimes used for calibration have spectra slightly different from that of sunshine. Therefore exact calibration of reference cells is quite difficult and irradiance measurements with cells calibrated by different institutions or at different times may vary by up to 5%. Thus the basic accuracy of reference cells is lower than that of pyranometers. The output voltage of many reference cells is in the range of 20 to 100 mV at 1 kW/m^2.

In addition to their reasonable price, reference cells have several benefits. If the main concern is not the true efficiency of conversion of sunlight into electricity, but *how well a PV plant operates*, it is best to relate its output to that of a reference cell of the same type (i.e. match crystalline silicon arrays to crystalline silicon reference cells). This will minimize the effects of changes in spectrum and reflection of light at the module surface, two unavoidable influences not indicative of deficiencies in plant design or performance. Many reference cells also have a built-in sensor to measure cell temperature, which is usually close to the average array temperature if the cell is mounted in a way similar to the modules of the array. This may also be used to correct the (relatively low) temperature dependence of the reference cell.

For low-cost applications, silicon photodiode sensors may be used, but their accuracy is generally lower than that of reference cells. Errors encountered in practical operation may be up to 10% at 1 kW/m^2. Output voltages are in the same range as those of reference cells. Photodiodes may have a large temperature dependence and are affected by spectral changes.

All sensors mentioned above also show some time dependency. They should therefore be *recalibrated at regular intervals* (e.g. every two years), if high monitoring accuracy must be guaranteed.

Irradiance and ambient temperature sensors should be protected from snow and ice accumulation. When tilt angles are high, snow accumulation can be minimized by placing the plane-of-array sensor near a sheltered wall or under one of the eaves of a roof; note that this may block some light to the sensor and will not prevent wind from blowing snow onto the sensor. For reliable, accurate irradiance measurements in cold climates, especially during winter, some ventilation and heating of the sensor is necessary. When lower accuracy is acceptable, the heating and ventilation system can be omitted. If it is assumed that the sensors will be free of snow and ice whenever the PV modules are generating power, the operation of the snow-free system can still be investigated, though losses due to snow and ice coverage will remain unknown. One problem with this approach is that snow and ice tend to melt or slide off PV arrays faster than they melt off sensors.

Heating and ventilation sets for pyranometers are commercially available. Figure 14.1 shows that such a heating and ventilation set managed to keep a pyranometer free of snow during a heavy snowfall of about 25 cm. The continuous power consumption of an energy-saving set is at least 5–10 W per pyranometer, depending on the outside temperature. Sets that are not optimized to save energy use much more power – up to 80 W. Heating and ventilation sets can be very expensive. They are rarely available for reference cells or photodiodes, but custom development is possible. For example, a reference cell can be heated using resistive heating foil bonded to the back of the cell; this will affect the built-in temperature sensor and will probably need more power than ventilating and heating a pyranometer.

ENERGY CONSUMPTION OF MONITORING EQUIPMENT

The energy consumption of a good data logger is quite low (e.g. less than 1 W), but auxiliary equipment, such as isolation amplifiers, current sensors and modems, also needs some power. Nonetheless, if no heating and ventilation set is used, the energy consumption of a simple monitoring system can be kept fairly low.

A significant increase in power consumption is caused by pyranometer heating and ventilation. Thus *accurately monitoring* even a small PV system in a cold climate demands additional electrical power of 20–25 W (continuous), or a daily energy supply of 0.5–0.6 kWh/d. In addition to the energy of the normal loads, this energy must be supplied by the PV system (or a separate power supply for the monitoring system) and will therefore increase the size and cost of the PV array and battery at off-grid sites. On the other hand, if a thermal management system is used, this dissipated energy may contribute to heating the essential parts of the system.

ELECTROMAGNETIC COMPATIBILITY (EMC) CONSIDERATIONS

A PV system with DC/DC or DC/AC power converters switching at high frequencies must have sufficient internal filtering to prevent radio-frequency noise from causing malfunction of the monitoring equipment. Every wire entering the monitoring equipment may act as an antenna for such signals and for transient overvoltages caused by switching operations or atmospheric events (lightning strikes, magnetic storms and polar aurora). Therefore, appropriate measures against overvoltages and noise must be taken to ensure reliable operation. *Careful protection and filtering of each signal wire increases the reliability of the monitoring system considerably:*

- Use shielded signal cables and ground one end only; in extreme cases both ends can be grounded, *but only if ground loops will not occur.*
- Avoid ground loops by using isolation amplifiers and differential inputs (if the common-mode range of the data logger is sufficient) and ensure that there is only one grounding point for the system.
- Tie floating circuits or sensors to ground with a resistor, thereby reducing parasitic common-mode voltages that might otherwise cause erratic behaviour.
- Use surge arrestors and filters on signal input lines.
- Ensure that power and signal lines are physically separated and do not share a conduit or raceway.
- Reduce internal noise from DC/DC or DC/AC converters by additional filtering of the power lines.

STANDARD DATA PRESENTATION FOR ASSESSMENT OF SYSTEM PERFORMANCE

Comparing the performance of different PV plants at different locations is complicated because the output of the array depends on the design of the system, as well as on the insolation and temperature at the site. Standard methods for analysis and presentation of monitoring data permit comparisons and have been introduced in the JRC Ispra *Guidelines*[1] and extended and improved by HTA Burgdorf.[5]

The total daily in-plane irradiation can be calculated by integrating the irradiance in the plane of the array over the day:

$$H_I = \int G_I dt.$$

By dividing this value by the standard test condition (STC) irradiance level of 1 kW/m², G_{STC}, the reference yield, Y_r, is obtained:

- *Reference yield* $Y_r = H_I/G_{STC}$: the number of hours per day that the sun would have to shine with a power of 1 kW/m² in order to equal the measured irradiation for the day, H_I; this is equivalent to the peak sun hours.
- *Normalized measures* of array and system performance are found by dividing relevant energy measurements (in

kWh) by the nominal PV array power at STC (in kW_p):

- *Array yield* $Y_A = E_A/P_O$: the number of hours per day that the PV array would have to operate at its STC power output, P_O, in order to generate the actual daily DC energy output of the array, E_A.
- *Final yield* $Y_f = E_{use,PV}/P_O$: the number of hours per day that the PV array would have to operate at its STC power output, P_O, in order to generate the energy output of the system that is actually used by a load or fed to the grid, $E_{use,PV}$.
- Capture losses $L_C = Y_r - Y_A$: the energy losses, expressed in hours per day of PV plant operation at STC power output, caused by cell temperatures higher than 25°C, losses in wiring and series diodes, poor module performance at low irradiance, partial shading, snow and ice coverage, module mismatch, operation of the array at a voltage other than its maximum power point, inverter failures (grid-tied systems) and/or batteries unable to accept further charge (in stand-alone systems). Note that when the cell temperature, T_C, is measured, temperature-dependent capture losses, L_{CT}, can be calculated and subtracted from the capture losses. The remaining losses, L_{CM}, permit much easier assessment of system performance and help identify problems.[5]
- *System losses* $L_S = Y_f - Y_A$: the energy losses due to inverter inefficiencies and/or battery storage losses, expressed in hours per day of PV plant operation at STC power output.
- *Performance ratio* PR = Y_f/Y_r: the ratio of useful energy produced by the PV system to the energy production of an ideal, lossless PV plant with 25°C cell temperature and the same irradiation. This gives a good indication how much of the theoretically available PV energy has actually been used.

The use of these normalized measures for a grid-tied system is discussed in the next section. In a stand-alone system, the most critical component is the battery. Plotting battery voltage (per cell) versus time gives a good first indication of system operation, especially when the time and size of discharge currents are known. For a more detailed analysis other, more specific diagrams (e.g. scatter diagrams showing mean battery voltage versus normalized charge/discharge current or graphs showing the frequency distribution of mean battery voltage) can be used.[6]

CASE STUDY

The *highest grid-connected PV plant in the world*, sited at Jungfraujoch (3,454 m above sea level) in the Swiss Alps, was designed and built by HTA Burgdorf during the summer and fall of 1993 (see *Case Studies*). This 1.13 kW_p system has been monitored since 27 October 1993. Despite the stresses imposed on all components by the alpine climate, the PV plant and monitoring system have operated nearly flawlessly.

Operating experiences

While the data acquisition system has been perfectly reliable, the power supply of the pyranometer ventilation system proved to be undersized and failed after only one month of operation; between December 1993 and June 1994, the pyranometers were covered by snow or ice for a period of several hours on several occasions. After replacement by a stronger power supply the ventilation and heating system performed as expected.

Two reasons for the success of the monitoring system are *frequent data analysis* and *redundancy*. In the Jungfraujoch installation, data is transmitted by modem via telephone lines to HTA Burgdorf daily, permitting regular verification of correct system operation. Sensors that are prone to failure are backed up by other sensors ('redundancy'). The pyranometer is checked against a reference cell and an AC energy counter is used to verify that AC power measurements are correct. Thus, when in February 1994 the AC power meter suddenly started exhibiting an error of 2%, the error was soon detected and the power meter replaced.

Sample data analysis

A convenient way of illustrating the efficiency with which a PV system uses the available PV power at a given location is a bar graph showing daily array and reference yields, Y_A and Y_r, for each day of a month or for each month of a year.[1] A similar but more sophisticated graph can also be created using Y_f, Y_A and Y_r, together with the performance ratio,[5] as illustrated in Figure 14.2. This shows the operation of the Jungfraujoch PV system, as revealed by monitoring, for each month of 1994. The vertical scale is hours per day. Thus, for January the reference yield Y_r was just above 3 hours per day, indicating that on an average day of January there were three peak sun hours of sunshine at the site. The array yield Y_A is slightly below 3, with the difference between the reference yield and the array yield attributable to losses caused by non-ideal operation of the array, L_C. Inverter conversion losses, L_S, further reduce the useful output of the system, Y_f, to the equivalent of 2.6 hours per day of operation of an ideal (lossless) PV system. The '85' at the top of the bar is the performance ratio for the month, showing that the actual useful output of the system is 85% of that of the ideal system (i.e. 2.6/3.05 = 85%).

The monthly performance ratios vary from 70% to 88%, with an annual average of 81.8%. Lower than average performance ratios from April through June were due to

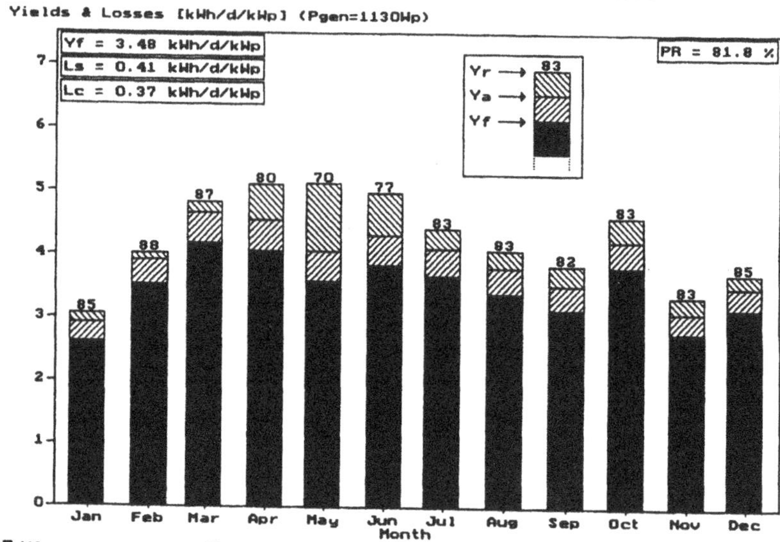

Figure 14.2. Monitored operation of the Jungfraujoch grid-connected PV system in 1994

snow accumulation on one of the two subarrays lasting several days. The capture losses L_C are generally lower than average during winter, probably due to lower array temperatures, which improve efficiency. System losses L_S are fairly constant, reflecting the fairly constant inverter efficiency, which averaged 90%. Note that 48% of the electricity for the year was generated in the winter months of October through March.

REFERENCES

1. JRC Ispra (1993). *Guidelines for the Assessment of PV Plants.* Document A: *Photovoltaic System Monitoring*; Document B: *Analysis and Presentation of Monitoring Data.* Ispra, Italy: Joint Research Centre of the Commission of the European Communities.
2. Wardle DI, Dahlgren L, Dehne K, Liedquist L, McArthur LJB, Miyake Y, Motschka O, Velds CA, Wells CV (1996). *Improved Measurement of Solar Irradiance by Means of Detailed Pyranometer Characterisation.* International Energy Agency Report IEA-SHCP-9C-2. Downsview, Ontario: Atmospheric Environment Service.
3. Imamura MS, Helm P, Palz W (1992). *Photovoltaic System Technology - A European Handbook.* Bedford: HS Stephens & Associates.
4. McArthur LJB, Dahlgren L, Dehne K, Hämäläinen M, Leidquist L, Maxwell G, Wells CV, Wardle DI (1995). *Using Pyranometers in Tests of Solar Energy Converters: Step-by-Step Instructions.* IEA Solar Heating and Cooling Programme Subtask 9F Report. Downsview, Ontario: Atmospheric Environment Service.
5. Haeberlin H, Beutler C (1995). Normalized representation of energy and power for analysis of performance and on-line error detection in PV Systems. *Proceedings of the 13th EU PV Conference, Nice.* HS Stephens & Associates, Bedford, UK.
6. Sauer DU, Bopp G, Bächler M, Höhe W, Jossen A, Sprau P, Willer B, Wollny M (1997). What happens to batteries in PV systems or do we need one special battery for solar applications? *Proceedings of the 14th EU PV Conference, Barcelona.* Published on behalf of WIP, Munich by HS Stephens & Associates, Bedford, UK.

FURTHER READING

International Electrotechnical Commission (1997). *Final Draft International Standard IEC 61724: Photovoltaic System Performance Monitoring – Guidelines for Measurement, Data Exchange, and Analysis.* Geneva: International Electrotechnical Commission.

Thevenard D, Ross M, Howell G (1998). A checklist for PV system monitoring. *Renewable Energy Technologies in Cold Climates '98, Montréal, 4-6 May.* Solar Energy Society of Canada, Inc., Ottawa, Ontario, Canada.

V Case studies

PV power for remote area schools in Argentina

THE APPLICATION

There are over 150 primary schools located in the remote and rural areas of the south-western Argentinean Province of Neuquén. Teachers are recruited from other parts of the country and are housed with their families in the school buildings. Of the 112 schools for which information is available, 34 are grid-connected, 42 have gas- or diesel-powered electrical generators, 27 use photovoltaic systems as their only source of power, one uses a wind-powered electrical generator and eight have no electricity at all. Although all of these systems require some level of maintenance throughout their service life, which the school teachers are sometimes able to undertake, frequent breakdowns of the gas and diesel generators result in school closures and lost time for the children. It is because of these problems that the provincial education ministry and power utility (Ente Provincial de Energia del Neuquén – EPEN) explored other means of electrical generation, primarily the installation of autonomous PV systems.

These generating systems are designed to provide electricity for the basic needs of the schools, primarily for lighting and for operating small electrical appliances such as communications devices, short-wave and AM/FM radios, VCRs and television. Wood fuel is the principal source of energy for cooking, water and space heating. Prior to the installation of the photovoltaic system, natural light was the only source of lighting and class hours varied from season to season according to the availability of natural light. A two-way radio powered by DC batteries is also available for emergency communication. The batteries for the radio are replaced periodically because of a lack of proper maintenance during the school holidays.

SITE AND CLIMATE

The picture above shows Escuela #306 Nahuel Mapi Arriba, which is located in the Province of Neuquén approximately 430 km south-west of the city of Neuquén, on the Auco Pan Native Reservation. The school is a two-room building, the largest being the main classroom, adjoining the living quarters of the teacher. It is located in a remote, mountainous region, 60 km from the electric grid, accessed by narrow, winding, dirt and gravel roads that are susceptible to washouts during rainstorms.

The school does not operate during the winter season as the snowfalls can be excessive in that region. The mean daytime temperatures at this location can reach as high as 33°C in the summer months from October to March and as low as –17°C in the winter months from April through October. Night-time temperatures consistently fall to near or below freezing throughout the year.

ECONOMICS AND ALTERNATIVES

Eight solar photovoltaic panels (each nominally 51 W, 3 A, 17 V) were provided by Darentek Corporation (Ottawa, Ontario, Canada) for a total cost of 2130 USD in September 1993. At the same time, three Solpac system controllers, provided by Reonac Systems (Ville St-Laurent, Quebec, Canada), were installed at 110 USD each. The local utility provided the balance of the systems and the installation labour.

The principal alternative to PV is the use of fossil-fuel-powered generators. As has been indicated, these units are frequently out of service as they are very difficult to maintain and operate properly, given the relative inaccessibility of the sites. Until the schools can become grid-connected, and this seems unlikely in the near future in some cases because of the rather low density of population in these rural mountainous areas, photovoltaic systems offer a meaningful option, provided they are well designed, installed and operated – taking into account all of the human problems of training.

LESSONS LEARNED

Notwithstanding the basic simplicity of a solar photovoltaic system, the remoteness of the sites where these schools are located, the inexperience of the staff operating the systems and the lack of training contribute to problems with maintenance. As a result, a Maintenance Manual has been written that details, in a step-by-step manner, the optimum procedures for system maintenance.

It is essential that the school caretakers be involved in the use and maintenance of the systems, given the relatively high turnover of the teaching staff. Steps have been taken by the local utility to ensure that its local offices verify the system maintenance. It has been found through experience that the batteries for the different systems should be changed every three years.

Other lessons learned are that a thorough preliminary analysis of the energy and logistics problems at the school is essential in order to achieve optimum performance from both technical and operational points of view.

In addition, educational aids and demonstration workshops have been developed to improve the understanding of the equipment. In general, the design of the photovoltaic and battery systems works quite satisfactorily. Most problems occur with its maintenance and with the training of the persons resident in these isolated locations.

The photovoltaic system provides the opportunity to make full use of television and video educational aids for the students. It is essential, of course, that adequate educational material in these formats be made available in order for this to happen.

In general, simple robust PV systems, with no moving parts, are less susceptible to breakdowns. Moreover, given that these systems are properly used and maintained, the savings in fuel and personnel costs justify the large initial investment.

Equipment acquired should be rugged, have low maintenance and be easy to operate. It is important to expand the package beyond just the photovoltaic system components to include training manuals, training courses, systems management and equipment maintenance. This package could enhance the standard of education in the schools, as well as offer a greater degree of comfort and well-being to the teachers and their families.

SYNOPSIS

Location	Neuquén, Argentina
Latitude	39.6°S
Altitude	400 m
Daily insolation on the horizontal:	
Average	4.05 kWh/m²/day
Worst month	1.16 kWh/m²/day (June)
Daily insolation on the plane of array	N/A
Air temperature:	
Average	20°C
Worst month	–14°C (July)
Period of operation	8–9 months/year
Installed PV power	424 Wp
Orientation	North
Inclination	45°
Storage capacity	12V, 660 Ah
Back-up power system	None utilized
Inverter size	None utilized
Cost of system	2800 USD

THE POTENTIAL

In Argentina, there are thousands of rural schools that are not connected to the grid. While this number is diminishing, there is great potential for photovoltaic systems to meet the basic electrical requirements of the remaining schools. Many of these schools are located in cold regions.

INFORMATION

More information can be obtained from:

Thomas A. Lawand/Josef Ayoub, Brace Research Institute,
PO Box 900, Macdonald Campus, McGill University,
Ste. Anne de Bellevue, Quebec, H9X 3V9, Canada
Tel. +1 514 398 7833 Fax +1 514 398 7767
Email ae12@musica.mcgill.ca

ACKNOWLEDGEMENTS

The Brace Research Institute would like to acknowledge the collaboration and assistance of the Ente Provincial de Energia del Neuquén (EPEN), Neuquén, Argentina and the Comision Nacional de Energia Atomica (CNEA), Buenos Aires, Argentina. The participation of the Institute, and the acquisition of components for the school cited in the example, is due to a grant from the Canadian International Development Agency in Ottawa, Canada.

PV power for monitoring natural gas wells in Canada

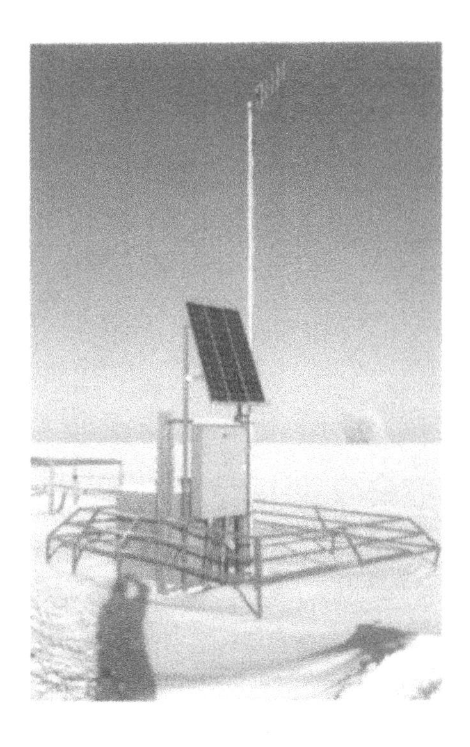

THE APPLICATION

Thousands of natural gas well heads are scattered around the countryside of central Alberta, Canada. In order to optimize the operation of these wells, gas companies install electronic equipment that monitors and, in some cases, controls the gas flows. This electronic equipment, generally known as a SCADA system, keeps in contact with a central pumping station by means of a radio, which receives and transmits signals for about one second each minute. The SCADA equipment requires a small but reliable supply of electricity.

PanCanadian Petroleum, one of the largest Canadian oil and gas companies, uses a photovoltaic system at each of its well heads as the sole source of power for the SCADA equipment. According to its specifications, this equipment draws about 11 W of power. Each system consists of 165 W_p of PV; tubular, lead–acid batteries, nominally 12V and 740 Ah, and a charge controller. The batteries are contained in an insulated enclosure, which helps maintain reasonable battery temperatures during the long, cold winter. Since the SCADA equipment must be operational at all times, the PV array and battery are conservatively sized.

SITE AND CLIMATE

The well head in the picture (above right) is situated on land leased from a farmer. There is a gravel access road to the site, which is 1.5 km from the electric grid. The land surrounding the site is used for livestock or crops.

The mean daily global radiation on a horizontal surface during December is about 0.8 peak sun hours (kWh/m^2). The coldest month is January, with a mean daily temperature of –14 to –18°C and extreme minimum temperatures of –43 to –52°C.

ECONOMICS AND ALTERNATIVES

The purchase price of the PV system is approximately 3500 USD, which is 15% to 20% of the cost of the entire SCADA system. The PV modules account for 30% of the initial cost, the batteries 30%, the insulated battery enclosure 15% and the controller 2%; the remainder is attributable to installation. If it is assumed that the system has a 20 year lifetime, the life-cycle cost is about 3000 USD, with taxation benefits due to depreciation responsible for the decrease in cost. The batteries are projected to last eight to ten years; if it is assumed that they are replaced once, their portion of the life-cycle cost is around 40%.

For most of these installations, the PV system is the least-cost option and thus has immediate payback. In the past, thermoelectric generators have been used to power many SCADA units. However, at many of the PanCanadian sites, standard thermoelectric generators cannot be used since they require purely gaseous fuel and the wells produce a mixed-phase mixture of natural gas.

While their purchase price, approximately 2400 USD, is lower than that of PV, thermoelectric generators cannot be placed within 25 m of the well head – a considerable inconvenience – owing to safety regulations. The load is too small to be met cost-effectively by a conventional gasoline or diesel generator and the maintenance requirements of the generator would be much higher than those of PV.

Grid extension is prohibitively expensive, susceptible to outages and lacks PV's modularity and simplicity of

PanCanadian employee inspecting the battery bank contained in an insulated enclosure

installation, which permit the power system to be moved to new well heads when the existing well heads are decommissioned. The wind resource varies from site to site; PanCanadian wanted to minimize administration costs by standardizing a single system configuration for all sites, which would not have been possible with wind turbines.

SYNOPSIS

Location	Alberta, Canada
Latitude	54°N
Altitude	1000 m
Daily insolation on the horizontal:	
Average	3.5 kWh/m²/day
Worst month	0.8 kWh/m²/day (Dec)
Daily insolation on the plane of array:	
Average	4.5 kWh/m²/day
Worst month	3.0 kWh/m²/day (Dec)
Air temperature:	
Average	2 °C
Worst month	–18°C (Jan)
Minimum air temperature	–47°C
Period of operation	Year-round
Installed PV power	165 W_p
Orientation	South
Inclination	70°
Storage capacity	12V, 740 Ah
Back-up power system	None
Inverter size	None
Average daily demand	264 Wh
Cost of system	3500 USD
Year of installation	1993

LESSONS LEARNED

Approximately 600 systems have been installed since 1992. PanCanadian Petroleum is pleased with the system performance. It will continue to install PV systems as standard equipment at its well heads and it is looking at new applications for PV.

There have been only a small number of system failures, which have been caused by malfunctioning charge controllers. The battery electrolyte must be topped up at an interval ranging from once every several months to once every two years, depending on the operation of the charge controller. The battery state of charge is maintained above 80% year round at most sites in most years.

In 1996, an in-depth technical and economic study was conducted with the goal of optimizing the PV system. It was found that the load was considerably smaller than specified and that a significantly smaller array and battery bank could be used.

THE POTENTIAL

PanCanadian Petroleum commissions 50 to 100 new well heads each year and there are several other companies that are equally active in the Alberta gas fields. SCADA systems are used at oil and gas installations around the world; globally, there are probably thousands of SCADA systems installed in cold climates each year.

INFORMATION

More information can be obtained from:

Sylvain Martel, CANMET Energy Diversification Research Laboratory (CEDRL), Natural Resources Canada, 1615, Lionel-Boulet Blvd., C.P. 4800, Varennes, Quebec, J3X 1S6, Canada
Tel. +1 450 652 6747 Fax +1 450 652 5177
Email sylvain.martel@nrcan.gc.ca

ACKNOWLEDGEMENTS

CEDRL would like to acknowledge the work of the following parties in this project:

Charles Ngai and Peter Chang, PanCanadian Petroleum Ltd.
150-9th Avenue S.W., PO Box 2850, Calgary, Alberta, T2P 2S5, Canada
Dave Egles, Soltek Solar Energy Ltd.2-745 Vanalman Ave, Victoria, BC, V8Z 3B6, Canada

South Pole expedition

THE APPLICATION

A photovoltaic system was used to provide electricity for a South Pole expedition. This expedition saw two explorers cross a vast section of the Antarctic continent on skis and pulkas (long sleighs pulled by a person). Bernard Voyer and Thierry Pétry skied a distance of 1500 km from the edge to the centre of Antarctica, reaching the South Pole in 63 days.

The PV system was used to power a satellite telephone under extreme climatic conditions. The PV system also served as a back-up for the lithium batteries for a video camera.

The NEC satellite telephone has its own internal Ni–Cad battery of 5 Ah capacity. This permits a maximum autonomy of half an hour without recharge. Because of the dangers inherent in such an expedition, it was originally planned to provide two independent power systems to the explorers, with two modules, two lead–acid batteries and a set of interchangeable cables, which could be used for all the power needs of the explorers. However, because of weight constraints, only one 9.5 Ah Sonnenschein lead–acid battery was taken; it was intended that this battery would be used in parallel with the telephone battery. After five days of operation, however, it was decided that only the satellite telephone battery would be used and it was charged directly by the photovoltaic panels.

The total energy demand was calculated to be 2.1 kWh for the duration of the expedition or about 35 Wh/day. Since it was estimated that the average daily insolation was going to be higher than 6 kWh/m^2, it was calculated that a 10 W_p module would be sufficient. Two UniSolar 5.5 W_p modules were chosen for their flexibility and durability. Mounted on top of the pulkas, these charged the batteries 24 hours a day.

An analogue voltmeter completed the equipment. This voltmeter permitted the explorers to check quickly and reliably that everything was working correctly. The entire system weighed 5 kg including the lead–acid battery.

SITE AND CLIMATE

The two explorers, Bernard Voyer and Thierry Petry, were the first North Americans to reach the South Pole unassisted, on skis. Each had to pull a pulka of 170 kg over an uneven surface of stratified ice swells. Sunlight was available 24 hours a day during their expedition. During sunny periods, the daily insolation reached 9 kWh/m^2. During white-out periods, the daily insolation was estimated at 6 kWh/m^2. During snow storms, the daily insolation was not more than 1.8 kWh/m^2. The record wind speed in Antarctica is 320 km/h while the record lowest temperature is –89.2°C.

ECONOMICS AND ALTERNATIVES

Primary lithium batteries are generally used on this kind of expedition. It was estimated that twenty 10 Ah lithium batteries would be needed, for a total cost of 4200 USD and a weight of 12.5 kg. This alternative was therefore rejected because of its cost and weight.

Another alternative was to carry a small gasoline generator. The smallest, lightest generator available was a 300 W generator weighing 18 kg. Apart from the weight of the generator, it would have to be carried and handled. Fumes and noise from the generator would be unpleasant and starting a generator in a cold environment would be difficult. This alternative was rejected because it was deemed impractical.

Solar photovoltaic energy represented the most economical, most reliable and most practical means of generating electricity for such an expedition.

LESSONS LEARNED

The PV systems were originally designed to power the following equipment:

- satellite telephone: 60 W (8 Wh/day)
- portable computer: 42 W (11 Wh/day)
- argo positioning system: 0.5 W (12 Wh/day)
- video camera: 20 W (4 Wh/day)

The total load was calculated to be a maximum of 35 Wh/day or 2.1 kWh for about 60 days of travel.

As it turned out, the explorers did not take the portable computer and Argo positioning system. However, they did use the satellite telephone more than anticipated, largely because there was sufficient power to do so. Thus, the load averaged 49 Wh/day. On dark days, the phone was used less; on sunny days, the explorers made more calls, including live radio interviews, giving them better public exposure.

This last-minute change in design underlines the importance of keeping the PV power equipment simple and flexible enough that its use can always be modified. In addition, extra use of communications equipment should be provided for, since public exposure is key to the overall success of such an expedition.

THE POTENTIAL

Photovoltaic systems are often used in fixed autonomous applications such as telecommunication towers, water pumping or remote dwellings. The development of flexible, durable modules and portable low-power equipment has created a new market for photovoltaics. Environergie has equipped many adventure expeditions with solar power. Since these expeditions are usually in extreme environmental conditions, the equipment must be reliable and adaptable. Photovoltaic systems have been used in expeditions going to the North Pole, the South Pole and Mount Everest. Photovoltaics can supply with extreme reliability the needs of these expeditions.

SYNOPSIS

Location	Antarctica
Latitude	78°35′ to 90°S
Altitude	260 m to 2800 m
Daily insolation on the horizontal:	NA
Daily insolation on the plane of array:	
Average	6.8 kWh/m²/day
Worst day	1.8 kWh/m²/day
Air temperature (Dec–Jan):	–15 to –33°C
Period of operation	9 November 1995 to 12 January 1996
Installed PV power	2×5.5 W_p
Orientation	Approximately north
Inclination	Variable between 10° and 70°
Storage capacity:	
Lead–acid	12V, 9 Ah
Ni–Cad	12V, 5 Ah
Back-up power system	None
Inverter size	None
Average daily demand	49 Wh
Cost of system	420 USD
Year of operation	1995–1996

DETAIL OF SYSTEM

The photograph below shows the modules in position on a pulka.

INFORMATION

More information can be obtained from:

Clément Bergeron, Environergie Énergie Solaire, 561 rue Pacifique
Lac-St-Charles, Quebec, G3G 1W6, Canada
Tel. +1 418 849 5221 Fax +1 418 849 5221
Email envirorj@clic.net

ACKNOWLEDGEMENTS

Environergie would like to thank Bernard Voyer and Thierry Pétry for their collaboration in keeping a daily usage log and Jimmy Royer of Solener Inc., Quebec, Canada for his assistance in designing the PV system for the expedition.

Diesel grid-connected PV system in Canada's Arctic

THE APPLICATION

In July 1995, a 3.2 kW_p photovoltaic power system was installed on the southern facade of the Nunatta campus building of the Nunavut Arctic College in Iqaluit, Northwest Territories, in the north of Canada. This system is connected to the local community grid, which is powered by large diesel generators.

The project had two principal objectives: first, to publicize photovoltaics as a viable power source in the Canadian Arctic and, second, to document the long-term performance of a large grid-tied photovoltaic system in the Northwest Territories.

When the economics of such systems are evaluated, hard data from an actual northern installation is more convincing than simulations alone or data based on southern installations. The Iqaluit system is meant to provide this hard data, as well as reveal any special problems associated with northern grid-tied systems.

The modules, which have a total area of about 25 m^2, are placed vertically on the front facade of the college, which faces 30° west of south. The PV array is connected to an inverter that transfers energy to the college's electric grid, reducing both its electrical consumption from the grid and the diesel consumption of the community.

SITE AND CLIMATE

Located at 63.8°N and 68.5°W, in the east of Canada's Northwest Territories, Iqaluit is only 300 km south of the Arctic Circle. The mean daily global radiation on a vertical surface (facing south) during December is 0.7 peak sun hours (kWh/m^2) and the temperature sometimes reaches a minimum of –45°C in February. The system is in operation year-round.

EXPERIENCE TO DATE

The system has been in operation since September 1995 and has proved to be very reliable to date. As is typical with PV grid-connected systems, no maintenance has been required since system start-up.

The array has proven to be very well integrated in the facade of the building; in fact many people did not even notice there were PV modules on the building.

The PV array has operated with an efficiency of 11.0% to 11.2% during the sunny, cold months of February through April. During summer, efficiency is only 9.0 to 9.7%, owing to the higher temperatures. During winter, efficiencies are around 8.5% to 10%, with the reduction attributed to low light levels.

The system has generated roughly 2 MWh of electricity per year. This is roughly 20% lower than suggested by simulations. This discrepancy has been attributed to module-mismatch losses, dust accumulation on the array and poor low light performance not accounted for in the simulation.

ECONOMICS AND ALTERNATIVES

For economic reasons stemming from their remoteness, over

300 communities in the north of Canada are not connected to a main electrical grid. The communities that have electricity have their own local grid. The majority of these local grids rely on diesel generators for their electricity.

Diesel fuel can be extremely expensive in these regions because of the high cost of transportation. As a result, the cost of electricity is very high and sometimes exceeds ten times the national average. Even though a photovoltaic system will generate very little power during winter, the reduction in fuel use during summer can make photovoltaic systems attractive.

This demonstration project shows that PV power systems can technically be alternatives to a diesel-powered generator. However, with an estimated cost of 1.50 USD per kWh for this demonstration PV system, compared to 0.20 USD per kWh for diesel generation in Iqaluit, the economics of such diesel–grid-tied PV systems in the Northwest Territories deters any extensive installations at this time. Reductions in the cost of PV technology, as well as eventual increases in cost for delivering fossil fuel, should permit PV to compete in the medium term. In many communities wind is also a promising alternative to fossil fuel.

LESSONS LEARNED

Mounting PV panels on the college's wall was relatively straightforward, taking only one day to complete. However, more resources than anticipated were required to organize and complete the project. The remoteness of the community significantly contributed to that, both in terms of time and cost of transportation.

One barrier to the implementation of major grid-tied photovoltaic systems in northern communities may be scepticism about the technology's ability to function in the north. This project helps overcome this barrier by proving that PV performs very well in the harsh conditions of the Canadian Arctic.

THE POTENTIAL

It is projected that grid-tied photovoltaic systems will become cost-effective in northern Canada before they do so in southern Canada. In Canada, there are more than 100 communities around or above the 60th parallel that are not connected to the North American electrical grid.

SYNOPSIS

Location	Iqaluit, Northwest Territories, Canada
Latitude	63.8°N
Altitude	40 m
Daily insolation on the horizontal:	
Average	2.6 kWh/m²/day
Worst month	0.1 kWh/m²/day (Dec)
Daily insolation on the plane of array:	
Average	3.2 kWh/m²/day
Worst month	0.7 kWh/m²/day (Dec)
Air temperature:	
Average	–10°C
Worst month	–26°C (Jan)
Minimum air temperature	–45°C
Period of operation	Year-round
Installed PV power	3.2 kW
Orientation	30°W of south
Inclination	90°
Storage capacity	None
Back-up power system	Grid
Inverter size	3 kW
Cost of system	70 000 USD
Year of installation	1995

If it is assumed that there will be moderate annual reductions in the cost of photovoltaic technology, such grid-tied photovoltaic systems is expected to be cost-effective in about 20 years.

INFORMATION

More information can be obtained from:

Sylvain Martel, CANMET Energy Diversification Research Laboratory, 1615, Lionel-Boulet Boulevard, C.P. 4800, Varennes, Québec, J3X 1S6, Canada
Tel. +1 450 652 6747 Fax +1 450 652 5177
Email sylvain.martel@nrcan.gc.ca

ACKNOWLEDGEMENTS

This demonstration was conducted under the aegis of the *Photovoltaics for the North* programme, which was supported by the Aurora Research Institute, the Nunavut Research Institute and the CANMET Energy Diversification Research Laboratory.

PV power for lighting in bus shelters

THE APPLICATION

This PV system is used for illumination in bus shelters. The lamp is controlled with daylight and infrared sensors such that it shines only when someone is in the shelter at night; this reduces electricity consumption. It is estimated that the 11 W fluorescent lamp operates about two hours a day, depending, of course, on bus and passenger frequencies.

The whole PV system has been mounted high up on a pole. This permits freedom in the location of the bus shelter, as the orientation and tilt can be optimized without influencing the design of the bus shelter. Furthermore, the high location gives much more power in winter when shading from nearby objects can be a major problem. It is hoped that this mounting arrangement will also deter thieves and vandals.

SITE AND CLIMATE

The bus shelters have been introduced in Gothenburg, Sweden, by the local bus company GLAB. The latitude of the city is 58°N. This city, close to the North Sea, with prevailing south-westerly winds heated by the gulf stream, has an average temperature in January just below 0°C, with extremes down to –20°C for several days or even weeks when the winds come from north or east. Therefore, the system has to be designed for very cold climates. Because of the high latitude and cloudy climate, the average daily insolation on a horizontal surface is only about 0.2 kWh/m^2 in December.

ECONOMICS AND ALTERNATIVES

The purchase price of the PV system is approximately 1300 USD, which is about 15% of the cost of the entire bus shelter. For bus stops in the countryside there is seldom grid electricity available and the extension cost is in the range of 15 000–30 000 USD/km and then the annual fixed cost for the connection is about 300 USD/year. Even in towns, grid extension is often not an alternative as digging trenches for cables, checking for existing underground cables and pipes and repairing paved surfaces may cost 6000 USD just for a few metres.

The PV option also has the big advantage that it can be easily moved to another location if there are changes in the demand for transport in an area.

The operating costs have been very high during the introduction phase as a result of theft and vandalism. Most of the original bus shelters with integrated PV modules have therefore been removed and just a few prototypes are now being tested with the design shown in the photograph below. So far the experience has been much better since this change. A stolen PV system is probably easy to sell in Gothenburg as it is close to the sea and many sailboats in the area use photovoltaics.

LESSONS LEARNED

Theft and vandalism can be a decisive problem for the introduction of PV in new applications. At high latitudes the sun appears low in the sky during winter, creating long shadows. This may limit where the bus shelters can be placed, especially in towns. A bus shelter can seldom be placed in a sunny location and facing south. Mounting of the PV system on a pole will help to overcome these problems as the risk of shading will be reduced and the PV module can be rotated and tilted independently of the orientation of the bus shelter itself.

THE POTENTIAL

The potential is very large as this solution is applicable to bus routes all over the world. Several companies are already active in this area, and not only in Sweden. This case study shows that photovoltaic-powered bus shelters can be used, albeit with some important design changes, in cold and not so sunny climates.

DETAIL OF SYSTEM

The picture shows the PV system with controller and battery on the top of the pole and the lamp with built-in sensor inside the shelter.

SYNOPSIS

Location	Gothenburg, Sweden
Latitude	58°
Altitude	100m
Daily insolation on the horizontal:	
Average	2.7 kWh/m²/day
Worst month	0.2 kWh/m²/day (Dec)
Daily insolation on the plane of array:	
Average	3.3 kWh/m²/day
Worst month	0.6 kWh/m²/day (Dec)
Air temperature:	
Average	7 °C
Worst month	–1°C (Jan)
Minimum air temperature	–25°C
Period of operation	Year-round
Installed PV Power	50 W_p
Orientation	South
Inclination	60–70°
Storage capacity	12V, 25 Ah
Back up power system	None
Inverter size	None
Average daily demand	20 Wh
Annual yield	Maximum 50 kWh
production, but only 10	kWh used for lighting
Performance ratio	0.1–0.2
Cost of system	1300 USD approximately for prototypes
Year of installation	1994

INFORMATION

More information can be obtained from:

Alf Hedman, Copyskylt AB, Box 105, S-26121
Billeberga, Sweden
Tel. +46 4184 314 00 Fax +46 4184 313 85

ACKNOWLEDGEMENTS

This presentation has been possible due to support and information from GLAB (Gothenburg local traffic company), ELFORSK (Research organization for the Electric Power Utilities in Sweden) and NUTEK (Funding organization for industrial development projects) in Sweden.

PV–H_2 pilot plant for experimental study

THE APPLICATION

Seasonal storage of solar energy is a challenging technical problem to which hydrogen technology may provide a feasible solution. In a photovoltaic–hydrogen (PV–H_2) system, excess photovoltaic electricity is fed to the electrolyser to produce hydrogen and oxygen from water. The hydrogen thus produced is stored over the season to be converted back to electricity in a fuel cell in the winter. Thus, hydrogen is used as a storage medium to shift solar energy from summer to winter.

The purpose of this pilot plant is to test the technical feasibility of the whole PV–H_2 concept and its separate system components. The system is not operated continuously, but different kinds of test runs are performed to get operating experiences in real conditions.

This particular PV–H_2 power system is able to meet a daily load of 1 kWh. It can easily be upgraded to meet a larger load. A certain economy of scale is attainable since, the bigger the hydrogen storage tank, the less it will cost per unit volume.

SITE AND CLIMATE

The PV–H_2 pilot plant is located at the solar energy testing site at the Helsinki University of Technology in Espoo, Finland. The average daily insolation at this site is 2.6 kWh/m^2/day. In July, the daily average is 5.8 kWh/m^2/day, and in December it is 0.2 kWh/m^2/day. The mean temperature is +5°C; varying from –6°C (January) to +17°C (July). The extreme minimum temperature is about –30°C.

OPERATING EXPERIENCE

Operating experience during the years 1992–96 has revealed much about the reliability and durability of PV–hydrogen technology. The main observations are as follows:

- The PV array and battery have operated smoothly and no particular problems have been observed.
- The electrochemical performance of the electrolyser has been good, but the gas-handling system has operated unreliably. The most common defects have been small leaks, poor reliability of electromagnetic valves and too sensitive safety limits of the automatic process control unit, occasionally causing unnecessary shut-downs.
- The phosphoric-acid fuel cell (PAFC) proved to be unsuitable for a small-scale self-sufficient system – firstly due to the open-end stack construction without a hydrogen recirculation loop, and secondly due to the high power consumption of the preheater. Therefore, a solid-polymer fuel cell (SPFC) seems to be more feasible for these systems.
- The control system has worked well and the direct coupling of the system components has been a good configuration. The use of the battery as a buffer storage between the PV array and the electrolyser, and also between the load and the fuel cell, has saved the electrolyser and fuel cell from rapid power variations.

The overall costs of the experimental PV–H_2 power system were approximately 100,000 USD. The seasonal storage system (electrolyser, hydrogen store and fuel cell) made up

80% of this. In the last ten years, the prices of hydrogen energy components have decreased remarkably and today the projected cost of such a system is approximately 40 000 USD.

LESSONS LEARNED

The reliability of hydrogen-based systems must be further improved. In particular, the lifetime of the electrolyser and fuel cell must be increased. Several systems should be operated for several years in order to get information on the ageing of the components and the overall system. The ageing of a single component may significantly affect the performance of the overall system. The thermal behaviour of remote systems must be studied to avoid overheating in summertime and excessive cooling in wintertime.

POTENTIAL

Potential applications for this kind of self-sufficient PV–H_2 power system include remote telecommunication repeater stations and remote village power, far away from electric grid and other services.

DETAIL OF THE SYSTEM

A schematic of the operation of the PV–H_2 plant is shown.

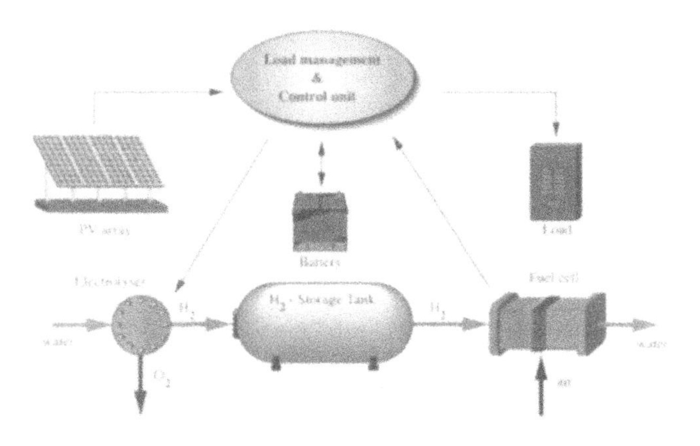

SYNOPSIS

Location	Espoo, Finland
Latitude	60°N
Daily insolation on the horizontal:	
Average	2.6 kWh/m²/day
Worst month	0.2 kWh/m²/day
Daily insolation on the plane of array:	
Average	3.1 kWh/m²/day
Worst month	0.2 kWh/m²/day
Temperature :	
Average	5 °C
Worst month	–6 °C
Minimum temperature	–30 °C
Period of operation	Year-round
Installed PV power	1.3 kW_p
Orientation	South
Inclination	45°
Storage Capacity:	
Battery	545 Ah (26 V)
Hydrogen storage	200 Nm³ (8 m³, 25 bar)
Cost of system	100 000 USD
Year of installation	1992

INFORMATION

More information can be obtained from:
Dr Peter Lund, Helsinki University of Technology
PO Box 2200
FIN-02015 HUT, Finland
Tel +358 9 451 3197 Fax +358 9 451 3195
Email peter.lund@hut.fi

ACKNOWLEDGEMENTS

This work has been funded by the Technical Development Centre of Finland through the national NEMO 2 – Advanced Energy Systems and Technologies Research Program.

BIBLIOGRAPHY

Kauranen P Lund, Vanhanen JP (1994). Development of a self-sufficient solar-hydrogen energy system. *International Journal of Hydrogen Energy*, 19, pp.99–106.

Vanhanen PS, Kauranen PD Lund, Manninen LM (1994). Simulation of solar hydrogen energy systems. *Solar Energy*, 53, pp.267–278.

PV-hybrid power systems for radio links in Greenland

THE APPLICATION

Tele Greenland A/S is responsible for all telecommunication in Greenland, including telephone, data transmission, maritime and air-to-ground radio systems and transmission of radio and TV programmes.

The country's 56 000 inhabitants are spread out along the coastal zones of this big island and the communication network is therefore based on radio links and satellite systems. The backbone of the network is a 1800 km radio link that covers the most densely populated part of the country – the southern part of the west coast. Satellite systems are used for long-distance transmissions going north–south and east–west, and side radio links connect the smallest settlements to the main systems. The radio links include 43 unmanned repeater sites placed on mountain tops along the coast line; PV–diesel systems are used for 30 sites, with power consumption in the range 400 to 1000 W, and PV–primary batteries are used for seven sites, with power consumption in the range 15 to 25 W. In a few cases PV–wind–diesel power systems have been used.

The title picture shows a PV–diesel powered site at 67°05′N, 53°49′W. Seen from left are:

- the storage building;
- an array with 16 × 48 W PV panels, and behind the array, the diesel genset building and fuel tank;
- the radio building with 24 × 48 W PV panels;
- two arrays with 16 × 48 W PV panels, and behind the arrays, the antenna tower.

SITES AND CLIMATE

There is no power grid for these repeater stations to connect to, so each must have its own power supply. The sites are accessible only by helicopter, and servicing and fuel transport for conventional fossil-fuel- or battery-powered systems is rather expensive. This is the main reason that Tele Greenland has chosen hybrid power systems.

Located above and below the Arctic circle, the sites are characterized by dark winters, long summer days, cool temperatures in the summer and extreme cold during the winter.

PRINCIPLES OF OPERATION

PV–diesel systems

The radio equipment is powered by a 48 V DC battery, to which the PV arrays and the diesel generators supply their energy.

When the PV arrays cannot supply enough energy to meet the load, the battery is allowed to drop to 65% of its nominal capacity. At this limit, one of the diesel generators gets a start signal. The diesel generator starts, tops up the battery charge and then stops.

In case of a start failure or other engine problems, the remaining 65% of the battery capacity represents spare capacity, which can supply the load for a couple of days. The number of days of battery autonomy is decided individually for each site, according to the accessibility of the site.

PV–primary battery systems

These are very simple and reliable systems, where the primary batteries take over in the dark winter months and supply the current to the load and float charge the Ni–Cd battery.

The number of primary battery cells depends on the location of the site and the desired interval between battery replacements.

ECONOMICS AND ALTERNATIVES

The total price for the PV–diesel hybrid system on the site shown in the picture is around 400 000 USD. This price includes all power equipment, the diesel building (24 m^2) with heat storage, installation work on site and helicopter transport.

The diesel part of the system is based on the equipment used at bigger sites where the diesel generators are run continuously. It includes two Lister Petter type TS2 diesel generators of 8 or 12 kVA.

The PV part, including all related equipment, represents an expense of 65 000 USD. Compared to continuous diesel-generator systems, the PV addition saves around 25 000 USD a year. The telecommunication link is also more reliable in that it relies on two independent power systems to operate.

The total price for the PV–primary battery system is around 120 000 USD. This price includes all power equipment, the building of area 9 m^2 for the Ni–Cd battery, the control panel and radio link. There is no heating system in this building. The primary batteries consist of 2000 Ah (nominal capacity) mercury-free SAFT AD 308 MF cells.

A typical installation for three years without service includes three battery sets, each of two 2000 Ah. The capacity is typically reduced by 35–50% at low temperatures. The batteries are placed outside the building in battery containers made of plywood and galvanized steel. The lifetime of the batteries after activation is three to four years. Experience with the batteries is good.

LESSONS LEARNED

Tele Greenland's use of PV energy started in 1985 and until now experience with the established systems and their cost savings has been good. Efforts are presently concentrated on improvement of the PV–diesel charge control, which could lead to a further reduction of the diesel runtime.

Today the total installed PV systems in Tele-Greenland's power systems represents 110 kW$_p$, or about 2 W$_p$ per inhabitant in Greenland.

SYNOPSIS

Latitude	From 61°N to 77°N
Altitude	From 400 to 1600 m above sea level
Insolation (horizontal)	0.2 to 3 kWh/m^2/day
Outside temperature	–40°C to +15°C
Wind speed	Up to 70 m/sec.

PV–diesel system site data

Power consumption:	
Typical	500 W
Maximum	1000 W
Nominal system voltage	48 V DC
Installed PV power:	
Typical	3400 W$_p$
Maximum	5700 W$_p$
Battery	lead–acid, valve regulated
Typical capacity	3000 Ah, max. 5000 Ah
Diesel generators	Two: 8 kW or 12 kW
Service interval	One visit per year
Cost of system	400 000 USD

PV–primary battery systems site data

Power consumption	15 to 25 W
Voltage limits	22 to 36 V
Installed PV power	300 to 800 W$_p$
Battery:	Nickel–cadmium
Capacity	450 Ah
Primary battery:	Air depolarized cells
Capacity	2000 Ah
Service interval	From one to three years
Cost of system	120 000 USD

INFORMATION

More information can be obtained from:

Viggo Gahrn, Group Manager, Infrastructure – Power Supply Systems, Tele Greenland A/S, Projects & Engineering, Thoravej 4 DK-2400 Copenhagen NV, Denmark
Tel. +45 38 34 22 44 Fax +45 38 34 63 65
Email vig@tele.gl

PV–hybrid system at the Klagenfurterhütte mountain hut

THE APPLICATION

In Austria there are many mountain huts, which have been built to accommodate visitors for dining and sleeping. Some of these huts are well equipped, having a kitchen, guest-rooms, showers and washrooms. There is usually a caretaker who provides services for the visitors.

Often it is not possible to connect these huts to the public grid. They are built away from main roads and power lines. They usually operate a diesel genset for electric power generation.

Increasingly, photovoltaic systems are used for these mountain huts. The Austrian electric utilities have more than ten years experience in this field and support such research projects. By installing the additional PV array, significant reductions in the operation of the conventional diesel system can be achieved. This results in reduced fuel costs and prevents pollution and noise.

The Klagenfurterhütte is located in the southern part of Austria at an altitude of 1663 m. There is no connection to the public grid. It is a well known hut, operated by the Österreichischer Alpenverein. The site is in operation throughout the year. During the summer (May to October) it receives guests on a continuous basis. During winter, it is open at weekends only.

The PV system at the Klagenfurterhütte is nominally 2544 W_p. It consists of 48 Siemens M55 modules, each with a peak power of 53 W_p. The modules are interconnected to form three subarrays. The generated power is then fed by means of three MPP-tracker controllers to a lead–acid battery. The nominal capacity of the battery is 36 kWh (12 Bären OPzS 2 V cells for a total of 24 V, 1500 Ah) providing four days of autonomy without recharge. The batteries are especially designed for PV operation and have an acid circulating pump to prevent stratification of the acid. This improves battery performance.

The system is completed by a diesel-powered 20 kW generator, which is used to charge the battery and power the loads when the PV system is unable to do so.

SITE AND CLIMATE

Located at 1663 m, the system experiences climatic conditions typical of the Austrian Alps. The minimum temperature at the hut is around –25°C, and the December insolation on the horizontal averages 0.5 kWh/m^2/day.

ECONOMICS AND ALTERNATIVES

The cost of the overall system (PV generator, distribution panel, cables, charge controller, charger, two inverters, 36 kWh lead–acid battery) including installation was about 47 000 USD.

The total power demand of all the loads is 5470 W with an average energy demand of 7756 Wh/day (341 Ah/day) in summer. The power demand is 1126 W, with an average energy requirement of 4080 Wh/day during the weekends in winter.

Using only a diesel generator, the fuel consumption would be more than four times higher, since the efficiency of a diesel generator is very low when powering a small load.

With the PV system, the hut is more pleasant, because the noisy diesel generator operates considerably less often. Only during periods with high power and energy demand is it necessary to start the diesel system, and then usually for a short time only.

LESSONS LEARNED

The hut's loads operate at the normal 220 V AC from the inverters, which transform the DC power from the battery to AC power for the loads. At first, only one inverter with 3000 W nominal output power was installed. The high output power is necessary to operate the dishwasher. However, problems occurred with small electronic devices, especially lights with electronic ballasts, because the original inverter produced a trapezoidal (modified square wave) waveform. It was then decided to install an additional 600 W inverter with a sinusoidal output voltage and use it only for the lights. The other inverter is used for the dishwasher and other kitchen and household appliances.

An extensive monitoring system has been installed to provide detailed data about the power generated by the solar and diesel systems and the electricity consumed. The average annual production is 1800 kWh by PV and 850 kWh by the diesel generator. The average energy demand is 1450 kWh/year. Thus 45% of the power generated is not being used. Part of this is accounted for by battery and inverter conversion inefficiencies, but two thirds of these losses are due to the stand-by power consumption of the inverter (141 kWh) and the monitoring system itself (262 kWh), both of which operate continuously. After the test period the extensive measurement equipment will be removed. This will reduce the DC losses considerably.

DETAIL OF SYSTEM

A schematic of the system operation is shown in the diagram.

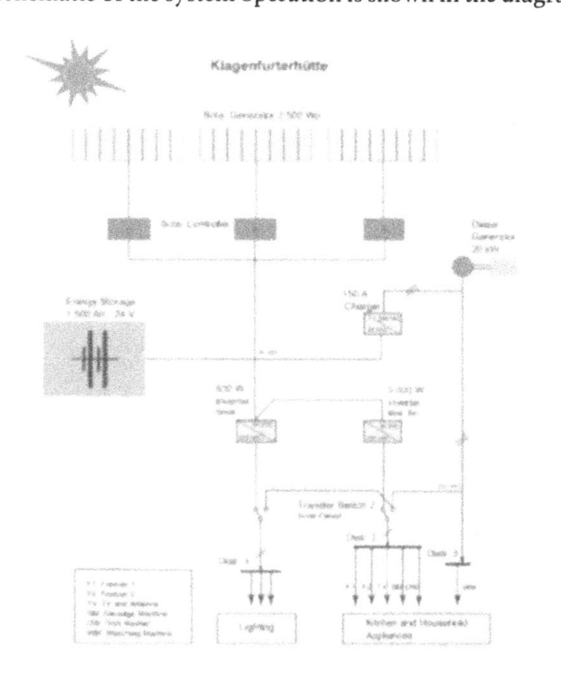

SYNOPSIS

Location	Austria, Bärental (near Klagenfurt)
Latitude	46.5°N
Altitude	1663 m
Daily insolation on the horizontal:	
Average	1.6 kWh
Worst month	0.5 kWh (Jan)
Horizontal	
Air temperature:	
Average	5 °C
Worst month	–4°C (Jan)
Minimum air temperature	–25°C approximately
Period of operation:	
May to October	Continuous
Winter	Weekends only
Year of installation	July 1993
Installed PV power	2544 W_p
Orientation	South
Inclination	40°
Storage capacity	24 V, 1500 Ah
Back-up power system	20 kW diesel engine
Inverter 1	Siemens SWR 600, 600 VA continuous
Inverter 2	VKL-Trapezoid, 3 000 VA continuous
Annual production:	
PV	1800 kWh
Diesel	850 kWh
Energy demand	1 450 kWh
Cost of system	47 100 USD

The diesel generator operates the washing machine directly when it is needed. When the batteries reach a certain depth of discharge, the diesel generator is started and recharges the batteries. During this time, the transfer switches let the generator power the loads.

INFORMATION

More information can be obtained from:

Ing. Josef Huber, Tauernkraft, Rainerstrasse 29, A-5020 Salzburg, Austria
Tel. +43 662 8682 22285 Fax +43 662 8682 12 22285
Email: huberjos2@verbund.co.at

BIBLIOGRAPHY

Huber J, Schauer G, Szeless A (1997). PV–hybrid system at Mountain Refuge Klagenfurter Hütte. *14th European. PV Solar Energy Conference, Barcelona (1997), Proceedings*, Vol. 2, pp. 2600–2603. Published on behalf of WIP, Munich by HS Stephens & Associates, Bedford, UK.

Photovoltaic plant of Kesselbach collecting works

THE APPLICATION

The Kesselbach collecting works consist of a Tyrolean dam with an intake gate preceded by a 35 m long desilter. The desilter is below ground with the overflow directed towards the Schlegeis-Rosshag power-plant pressure tunnel. A scouring gate for the desilter basin is placed at its lower end.

During the snow-melt period and during thunderstorms the greater risk of silt build-up in the desilter basin increases the risk of strongly polluted water entering the pressure tunnel, thus causing damage to the turbines at the Rosshag power plant. Electrification of the collecting works enables remote control of the gate and improved plant monitoring, thus avoiding energy losses. A photovoltaic plant was given preference over a conventional power-line supply for financial and environmental reasons.

SITE AND CLIMATE

The Kesselbach collecting works is situated in inaccessible terrain. It is located in the Ziller valley in the alpine part of Austria at an altitude of 1860 m. The average monthly temperature is between 10°C in July and –6°C in January. The minimum air temperature is below –20°C, with wind velocities up to 200 km/h. The duration of daylight is between 6 h (winter) and 16 h (summer).

The site is shaded between four hours (May) and five hours (December) a day. However, while the site does not receive a lot of direct sunlight during winter, reflections from snow and clouds do illuminate the modules. From November to April the site is covered with snow and is not accessible.

ECONOMICS AND ALTERNATIVES

The cost of the PV system, charge controller, battery, material and installation totalled 83 900 USD. The high price of the system is mainly due to its inaccessible alpine location. Materials had to be transported by helicopter. Because of the possibility of avalanches, it was necessary to make a very strong base for the two PV arrays using a steel structure.

A conventional supply network (power line) was not possible because of:

- the remote location of the intake gate;
- the higher price of the power line;
- the environmental difficulties of installing a conventional distribution line in the rocky terrain.

LESSONS LEARNED

The nominal efficiency of the modules (Arco Solar) is about 12% and the annual average efficiency is 10.4%. The annual average efficiency of the whole system is 9.3%.

The system was designed to meet energy demand in wintertime. Surplus production during summertime cannot be used. Therefore only 7.1% of the annual insolation is converted into electric energy to operate the devices of Kesselbachfassung. These values have not changed significantly since installation.

Except for replacement of the battery, the plant has been operating without any stoppages since installation. As a result of damage caused by low state-of-charge operation, the lead–acid gel battery had to be changed after four years of operation. The new battery (with acid circulation) operates without any problems.

To reduce the risk of interruption of the electric supply by avalanches, the modules are erected on two different sites. The two solar arrays are oriented towards the south at a slope of 60°; each consists of 16 modules. The arrays are mounted on galvanized iron frames. To avoid problems with high snow levels the lower frame edges are 2 m above the ground.

The photovoltaic plant has a maximum power output of about 1500 W at a rated voltage of 24 V. A battery plant with a total capacity of 1800 Ah can power the Kesselbach collecting works for at least two weeks of cloudy weather. The 24 V battery plant provides energy for the electric motor of a hydraulic unit for the intake gate and the direct drive of the scouring gate, as well as for measuring, control, remote-control and telecommunications equipment. The battery and all electronic devices are set up in a room inside the collecting works.

SYNOPSIS

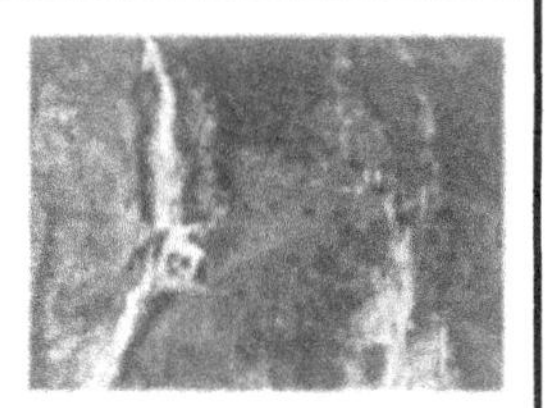

Location	Tyrol (Zillertal), Austria
Latitude	47°N
Altitude	1860 m
Daily insolation on the plane of array:	
Average	3.7 kWh/m²/day
Worst month	1.24 kWh/m²/day (Dec)
Air temperature:	
Average	2°C
Worst month	–6°C (Jan)
Minimum air temperature	–22°C
Period of operation	Year-round
Year of installation	1989
Installed PV power	1500 W_p
Orientation	South
Inclination	60°
Storage capacity	1800 Ah, 24 V
Back-up power system	None
Average daily demand:	
Summer	2400 Wh
Winter	1800 Wh
Inverter size	1000 VA, 230 V, 50 Hz
Annual production	1370 kWh
Performance ratio	56%
Cost of system	86 000 USD

INFORMATION

More information can be obtained from:

Dipl.-Ing. Dr Gerd Schauer, Verbundgesellschaft, Am Hof 6a
A-1010 Wien, Austria
Tel. +43 1 53113 52439 Fax +43 1 53113 52469
Email SchauerG@verbund.co.at

BIBLIOGRAPHY

Rabensteiner G, Huber J (1991). *Photovoltaik im Kraftwerksbetrieb* (Photovoltaic in Power Station Operation). ÖZE 44 H. 3, pp. 102–107.

Schauer G, Friess M, Korczak P, Szeless A (1998).10 Years operational experience of join utility-Siemens PV Projects in Austria. *2nd World Conference and Exhibition on Photovoltaic Solar Energy Conversion*, Vienna (1998) (in preparation: Session Ref. No. 1101, VD6/32).

Schauer G, Szeless A (1996). Nutzung der Sonnenenergie zur Stromerzeugung – österreichische Fallbeispiele (Use of Solar Energy for Electrical Energy Generation – Austrian Case Studies). e&i, 113. Jg. H. 6, pp.436–442.

Grid-connected PV plant on Jungfraujoch in the Swiss Alps

THE APPLICATION

The highest grid-connected PV plant in the world, at Jungfraujoch (46.5°N, 3454 meters above sea level), was planned and implemented by the Berner Fachhochschule, Hochschule für Technik und Architektur (HTA) Burgdorf during the summer and autumn of 1993. It has operated successfully with 100% availability of energy production and monitoring data since 27 October 1993. Energy production (especially winter energy) reached record values for a PV plant in central Europe.

The plant is a 1.1 kWp grid-connected PV system that is used for testing components and analysing performance of PV systems in high-altitude environments.

The picture shows a PV array and the irradiance sensors (heated and ventilated pyranometer and reference cell) on the outer wall of the research station at Jungfraujoch.

SITE AND CLIMATE

Wind loads encountered at this location are extremely high and, because of the quite frequent thunderstorms, lightning and overvoltage protection are very important issues. Since October 1993, the plant has survived the following high alpine stresses without any damage:

- *heavy storms* with wind speeds above 200 km/h, a very hard test for mechanical components and construction;
- *thunderstorms* with heavy lightning strikes, causing damage to other experiments at the research station;
- *irradiance peaks* with values up to 1720 W/m^2, causing high peak currents;
- *large temperature differences*: on a cold winter day, the drop in the temperature of a PV cell after sunset can exceed 40°C within 30 minutes; the total range of measured temperature so far is –29°C to +66°C.
- *Snow and ice covering the solar generator*: in spring, snow cover more than 3 m deep is possible. The level of snow cover depends not only on the amount of snow that has fallen, but also on the wind speed and wind direction during and after the snowfall. Sometimes energy production is also reduced by frost and partial shading from large icicles that hang from the roof.

LESSONS LEARNED

The figure (facing page) shows normalized monthly energy production (monthly energy production divided by PV generator size in kW$_p$) of the PV plant Jungfraujoch (3454 m) and a PV plant in Burgdorf (540 m), Switzerland.

As a result of the deep snow cover in spring 1994, one of the two PV arrays was completely covered for a few days, causing a remarkable drop in energy production. Therefore, energy production in spring (April to June) was lower than in the following years. Referenced to nominal PV generator power (1152 W$_p$), energy production was 1247 kWh/kW$_p$. During 1994, 48.0% of the PV system's output energy was produced during winter (October to March), which was roughly as expected. The inverter was 89.6% efficient on average.

In 1995, irradiation on the array plane increased and snow coverage in spring was considerably less than in 1994.

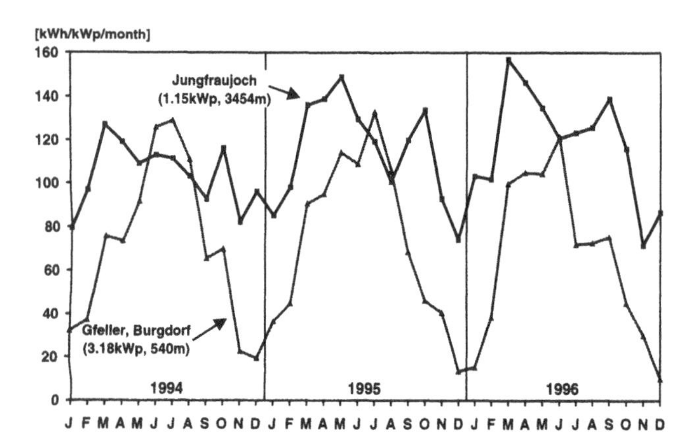

Referenced to nominal PV generator power, energy production was 1377 kWh/kW$_p$. The system produced 45.0% of its energy during winter and the mean value of inverter efficiency was 89.9%.

Snow coverage in spring 1996 was again rather low and irradiation on the array plane increased slightly compared to 1995. In July 1996 the string diodes in the array were eliminated and a new inverter with better efficiency was installed. Both measures improved performance compared to the respective months in earlier years. Energy production increased to 1426 kWh/kW$_p$. The energy produced during winter amounted to 44.6% of the year's total and the average inverter efficiency was 90.6%.

CONCLUSION

In all three years of monitored operation, energy production was relatively constant over the whole year, owing to the tilt angle of 90° and the high amount of sunshine in winter. Instead of the usual summer maximum and winter minimum (which can vary by a factor of ten in lower parts of Switzerland, as seen in the graph comparing Jungfraujoch and Burgdorf), two maxima per year, a higher one in spring (March, April or May) and a lower one in autumn (September or October), are observed. In summer, owing to the high albedo of the glacier in front of the PV array, much radiation is reflected onto the array, despite the high tilt angle of 90°. Summer energy production is therefore also remarkably high.

The only major operational problem encountered has been the temporary snow coverage occurring in spring. However, because of the tilt angle of 90°, this problem was not very serious. With the array installed higher off the ground (e.g. 5 to 7 m instead of only 3 m), this problem could probably be completely eliminated.

SYNOPSIS

Location	Jungfraujoch, Switzerland
Latitude	46.5°N
Altitude	3454 m
Daily insolation on the plane of array:	
Average	4.59 kWh/m²/day
Worst month	2.22 kWh/m²/day
Air temperature:	
Average	–6°C
Worst month	–13°C
Minimum air temperature	–29°C
Period of operation	Year-round
Year of installation	1993
Installed PV	1152 W$_p$ (24 Siemens M75)
Orientation:	
Array #1	12°W
Array #2	27°W
Inclination	90°
Storage capacity	None, grid-connected
Inverter	2.2 kVA, 230 V (Top Class 2500/4)
Average daily demand	Grid-connected
Annual yield	1592 kWh or 1384 kWh/kW$_p$
Performance ratio	82.5%
Cost of system	30 000 USD

INFORMATION

More information can be obtained from:

Heinrich Haeberlin, Berner Fachhochschule, Hochschule für Technik und Architektur (HTA) Burgdorf, Jicoweg 1, CH-3400 Burgdorf, Switzerland
Tel. +41 34 426 68 53 Fax +41 34 426 68 13

ACKNOWLEDGEMENTS

HTA Burgdorf is particularly grateful to the institutions that gave their financial support. The work described in this case study was funded by the Swiss Federal Office of Energy (SFOE/BEW), Berne and the Office for Water and Energy (WEA) of the canton of Berne. Construction of PV plant Jungfraujoch was sponsored by SFOE, VSE (Verband Schweizerischer Elektrizitätswerke), Siemens Solar (modules), Fabrimex Solar (inverter) and the Railways of the Jungfrau Region. Thanks go also to the International Foundation Scientific Stations Jungfraujoch and Gornergrat and the owner of the research station at Jungfraujoch, who permitted the use of its building for this project. HTA's PV activities in general are also supported by IBB, Burgdorf and EWB, Berne.

PV/TEG for Air Force ACMI systems

THE APPLICATION

Northern Power Systems (NPS) has designed and manufactured eight solar hybrid power systems for installation at remote locations in central Alaska under contract to Kollsman Systems Management.

The systems supply power to the remote stations required to support the US Air Force's Air Combat Maneuvering Instrumentation (ACMI) range in Alaska. The ACMI range is used to monitor Air Force training manoeuvres and gauge pilot performance. Sophisticated communications equipment relays the flight performance results from remote stations to the master station for evaluation and debriefing.

The power system of each ACMI range consist of a 760 W_p photovoltaic array coupled to an 80 W propane-powered thermoelectric generator (TEG). This powers a load made up of the ACMI itself, two lights, a fan and two heaters, which are used to heat up the battery enclosure and preheat the TEG fuel during cold spells. The winter load is evaluated at 20 W continuous, while the summer load is evaluated at 98 W continuous. The battery bank is composed of four Absolyte II GNB type 6-75A07 batteries connected in series to give 48 V. The battery capacity is 305 Ah at a 100 hour discharge rate, which is sized to power the communication loads through ten consecutive sunless days when fully charged.

The photovoltaic array is designed to power the site loads in the sunny spring, summer and fall months. The TEG powers the loads during Alaska's dark winter months. Both power systems charge the maintenance-free battery bank directly.

SITE AND CLIMATE

In winter, air temperature can be as low as –55°C and wind speed can be as high as 200 km/hr. An enclosure heater is used to prevent the battery from falling below a set temperature. In summer, the daily insolation can be very high, reaching 6.5 kWh/m^2/day. In order to avoid overheating of the enclosure, a temperature-controlled fan has also been incorporated into the design.

ECONOMICS AND ALTERNATIVES

A preliminary comparison of different alternatives has shown that cost of the PV/TEG/battery system with shelter was about 40 000 USD. A diesel system to do the same job would have to use dual 4.5 kW diesels, complete with rectifiers, battery shelter, etc., at a cost of 30 000 USD.

Operating cost for the TEG/PV system was estimated to be less than 2000 USD/year including helicopter time, because there is no maintenance and the only thing needed is propane at a cost of less than 250 USD/year.

Operating cost for the diesel system was estimated to be more than 10 000 USD/year. This is mainly because, with a cycle charge system, the diesel generators would probably have about 800 hours of run time per year, requiring two oil changes and refuelling. Furthermore, the shelter would have to be physically larger and would have to be heated, probably using a separate diesel space heater. However, we have estimated that operating cost would be dominated by the two service trips and helicopter time required.

From these preliminary calculations, it was decided to use the PV/TEG hybrid systems for all eight sites.

LESSONS LEARNED

NPS designed a superinsulated chamber to protect the power system and communication electronics from the extreme cold of Alaska's winter months. The units were assembled at NPS's Vermont facility. This factory preassembly and testing minimizes on-site work and ensure a highly reliable low-maintenance power system. The completed systems were transported by helicopter to the remote sites. Once dropped onto site foundations, the only significant installation tasks required were the erection of the solar array, the connection of the TEG to its fuel supply and system commissioning.

NPS also supplied its Gorillage™ galvanized steel foundation structures, designed to provide support for both the communication towers and equipment shelters. This unique foundation design eliminated remote-site concrete transportation and mixing, which is very expensive at sites not accessible by ground transport.

THE POTENTIAL

The potential for unattended low-power-consumption PV/TEG hybrid systems is tremendous. PV/TEG hybrid systems have proven to be extremely reliable in very harsh environments.

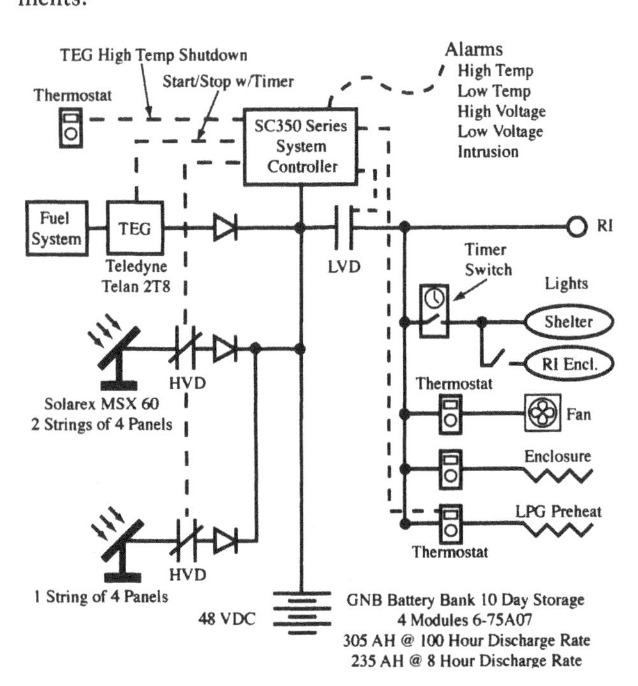

SYNOPSIS

Location	Sparraevohn, Alaska (eight sites)
Latitude	61°N
Altitude	760–1500 m
Daily insolation on the horizontal:	
Average	2.5 kWh/m²/day
Worst month	0.1 kWh/m²/day (Dec)
Daily insolation on the plane of array:	
Average	3.2 kWh/m²/day
Worst month	0.6 kWh/m²/day (Dec)
Air temperature:	
Average	–3.8°C
Worst average (Dec)	–23°C
Minimum air temperature	–51°C
Period of operation	Year-round
Installed PV power	760 W_p (12 MSX 60)
Orientation	South
Inclination	60°
Storage capacity	48 V DC, 305 Ah at 100 h
Back-up power system	80 W TEG with N_2 pressurization system
Average daily demand	98 W continuous
Winter demand	20 W continuous
Cost of system	30 000 USD
Year of installation	1990

DETAIL OF SYSTEM

As shown in the diagram (left), the power system consists of 12 Solarex MSX 60 modules connected in three strings of four modules to charge a 48 V battery. A SC350 Series System Controller detects when the battery needs charging. A timer switches the charging function exclusively to the TEG during winter's darkest month. The rest of the time the battery is recharged by the PV array.

INFORMATION

More information can be obtained from:

Jito Coleman, Northern Power Systems, One North Wind Road, PO Box 659, Moretown, VT 05660, USA
Tel. +1 802 496 2955 Fax +1 802 496 2953
Email: jcoleman@nwvt.com

ACKNOWLEDGEMENTS

Northern Power Systems wishes to thank Kollsman Systems Management Corp., which was the subcontractor for the Air Force, and the Alaskan Air Guard, who run the range.

PV–DG hybrid system for a remote residence

THE APPLICATION

Many remote places in Norway have been electrified by means of a diesel generator system (DG). Along the Norwegian coastline a wind–diesel generator system will often be a better solution and reduce the operation time of the diesel engine. In inland areas, where the wind conditions are not favourable for electricity generation, the solar radiation is good, at least in the summer season, and a PV system can make a significant contribution to the generation of electricity.

The objective of the system that was tested in this case was to provide power to a remote dwelling house as an alternative to a particularly expensive grid connection. The house was located at Venberget in Hedemark county in southern Norway. The house is about 20 km from the nearest grid. The electrical consumption of a typical house in this area is about 20 000 kWh/year.

In the wintertime, the contribution from the PV array is negligible. As a result, the average operating time of the genset is approximately six hours per day. On a fair day during the summer season the load is fully supported by the PV module. It is also expected that the PV modules will charge the batteries more fully than the genset and contribute to prolonging the lifetime of the batteries.

The system is fully automatic. The diesel genset and the PV/battery system alternate to supply the consumer with electricity. The genset is turned on or off automatically, depending on the load and the state of charge of the batteries. The genset starts when either the load is larger than a specified threshold (e.g. load greater than 1 kW) or the state of charge of the battery is below 50%.

The general maintenance of the system is partly done by the user and includes changing the oil in the diesel engine every 400 hours of operation.

SITE AND CLIMATE

At 60°N, the site experiences very low insolation levels during the winter and long days during the summer. The moderating influence of the Atlantic Ocean results in warmer temperatures than at most sites at this latitude, with minimum temperatures of around –25°C. The site is accessible by road.

ECONOMICS AND ALTERNATIVES

The price of the total system is approximately 60 000 USD, including a separate insulated building for the diesel genset. The recurring costs consist mainly of fuel consumption and replacement of the battery. For powering a single house, the system is competitive compared with a grid connection if the distance to the existing grid system is more than 3 km.

As stated already, a typical Norwegian dwelling house uses about 20 000 kWh of electricity annually. Most of the energy consumed is used for space heating and hot water. This part of the energy consumption can easily be replaced by other energy carriers, e.g. kerosene. For example, an electric stove may be replaced by a gas stove. In the current installation, electricity consumption has been reduced to approximately 4000 kWh/year. Further reduction is possible.

LESSONS LEARNED

The PV modules are arranged in nine groups of three modules connected in parallel. Each group of modules is connected to one battery module (three 6 V batteries in series) so that the charging of each battery module is done individually and independently of the charging of other modules. The intention is to counteract the differences between the battery modules. In contrast, the genset charges all the battery modules in series and this may not fully recharge all cells.

On bright summer days, when the insolation is good, the PV modules generate more energy than is consumed by the user. If the SOC of the battery system is 100%, the system will produce surplus energy that cannot be utilized and must be regarded as lost energy. To minimize the surplus energy, the system should anticipate future insolation. When the conditions are favourable, complete charging by the genset should not be necessary. Thus, the operating time of the genset can be reduced. To optimize the operation of the genset, a more sophisticated control system is needed, in which fuel consumption, proper charging of the batteries and other parameters can be taken into consideration. The project group plans to develop a programmable control system that would permit different operating strategies to be put into practice in order to gain operational experience.

SYNOPSIS

Location	Venberget, Hedemark, Norway
Latitude	60°N
Altitude	NA
Daily insolation on the horizontal:	
Average	3.0 kWh/m²/day
Worst month	0.35 kWh/m²/day (Jan)
Air temperature:	
Average	7°C
Worst month	–1°C (Jan)
Minimum air temperature	–25°C
Period of operation	Year-round
Year of installation	NA
Installed PV power	2.2 kW_p (27 MSX 83)
Orientation	South
Inclination	60°
Storage capacity	18 Varta GLS 6/300 batteries – 36 V DC, 1800 Ah
Back-up power system	Lister/Leroy TS2 diesel genset – 9.2 kVA
Energy demand	4000 kWh/year
Cost of system	60 000 USD

FUTURE PLANS

The possibility of substituting the DG with other heat and electricity generators is currently being evaluated. A Norwegian company plans to establish production of a micro Combined Heat and Power generator (CHP) using a Stirling engine as prime mover.

The operational characteristics of a CHP generator fit well in a hybrid system with solar energy, since they complement each other. The electricity consumption is not dependent on the outside temperature and is therefore unchanged throughout the year. In contrast, the heat consumption is dependent on the time of the year and has its maximum in the winter season and its minimum in the summer.

THE POTENTIAL

Despite the centralization that has occurred in Norway over the past decades, people are still living in rather remote areas. The electricity supply to such regions may be very expensive to maintain because of the long distance to a central grid system. A connection line will sometimes cost more than 100 000 USD to construct. Furthermore, when older lines have to be reconstructed owing to ageing and wear, alternative solutions based on local power-generation systems are being considered. It has been estimated that the number of houses where such local stand-alone power systems can be an economical solution is between 1000 and 3000, depending on the cost of the grid connection. In addition, there exist an unknown number of remote cottages and cabins, tourist lodges and other remote sites that require a high standard of energy services.

INFORMATION

More information can be obtained from:

Dr Alf Bjørseth, ScanWafer A/S, PO Box 290, Ørnesveien 3, 8160 Glomfjord, Norway.
Fax +47 75 71 90 13
Email alf.bjorseth@scanwafer.com

Appendices

A Solar radiation maps

Figures A.6, A.7, and A.8 are maps of the world with contours of equal insolation, given in cal/cm². The maps show average daily radiation on a horizontal surface during the months of June, September and December.

The data in these maps must be transformed from the horizontal to the plane of the array. The charts below can be used to estimate the plane of array radiation for fixed arrays facing the equator. For June and September, use Figures A.1 and A.2, respectively. For December, find the chart that contains conversion factors for the latitude of the site: Figure A.3 for latitudes of 30 to 50°N, Figure A.4 for latitudes of 55 and 60°, and Figure A.8 for 65°N. Above this latitude, there is essentially no sunshine in December, and the tilt factor is zero.

In Figures A.3 through A.5, representative tilt angles for the latitude have been given. There is a high tilt angle, which is roughly optimum for winter collection, a moderate angle, which is roughly optimum for year-round operation (and is sufficiently high to facilitate snow shedding in a cold climate), and a low tilt angle, which is roughly optimum for summer operation.

The tilt factor depends not only on the latitude, but also on the mix of diffuse and beam radiation. At cloudy sites, there is more diffuse radiation and tilt factors are lower. At clear sites, there is more beam radiation and tilt factors are higher. This effect is especially important during winter months and at high latitudes. For this reason, Figures A.3 through A.5 include insolation level as a variable.

Note that these tilt factors are for *monthly average* data and should not be applied to radiation on the horizontal for a single hour or a single day.

The tilt factors are prepared for the northern hemisphere. For the southern hemisphere, use Figure A.1 for December and Figures A.3, A.4 and A.5 for June.

To convert cal/cm² to other units, use the conversion factors in Table A.1.

Table A.1. Insolation conversion factors

To convert from . . .	to . . .	multiply by . . .
cal/cm²	kWh/m²	0.0116
cal/cm²	MJ/m²	0.0418
MJ/m²	kWh/m²	0.278
kWh/m²	MJ/m²	3.6

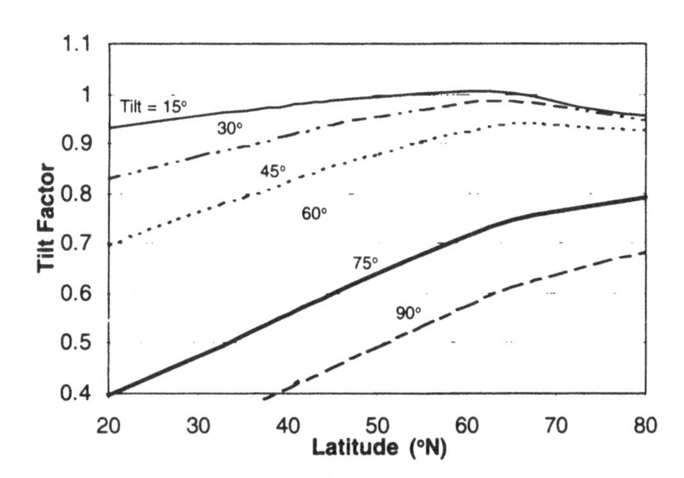

Figure A.1. Tilt factors for average daily radiation on the horizontal for June

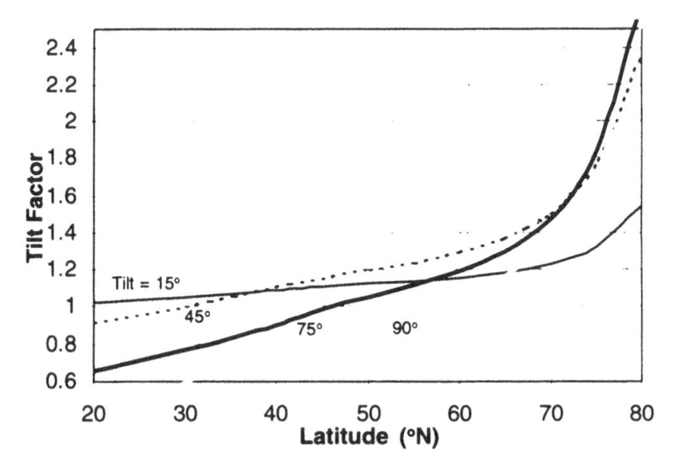

Figure A.2. Tilt factors for average daily radiation on the horizontal for September

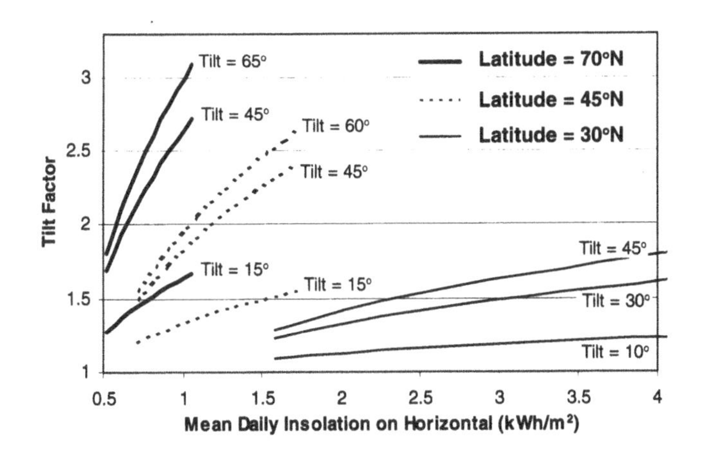

Figure A.3. Tilt factors for average daily radiation on the horizontal for December, latitudes 30° to 50°N

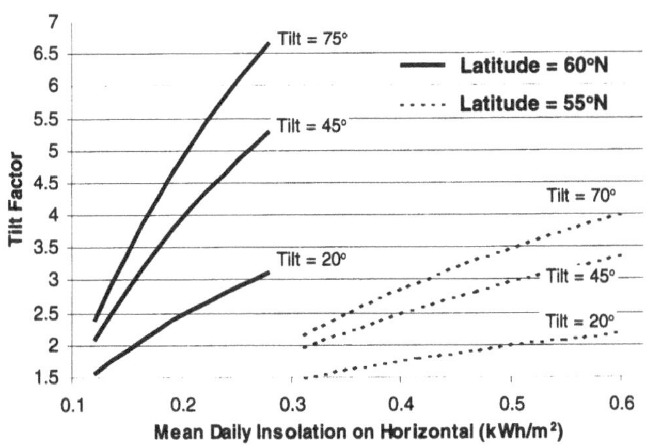

Figure A.4. Tilt factors for average daily radiation on the horizontal for December, latitudes 55° to 60°N

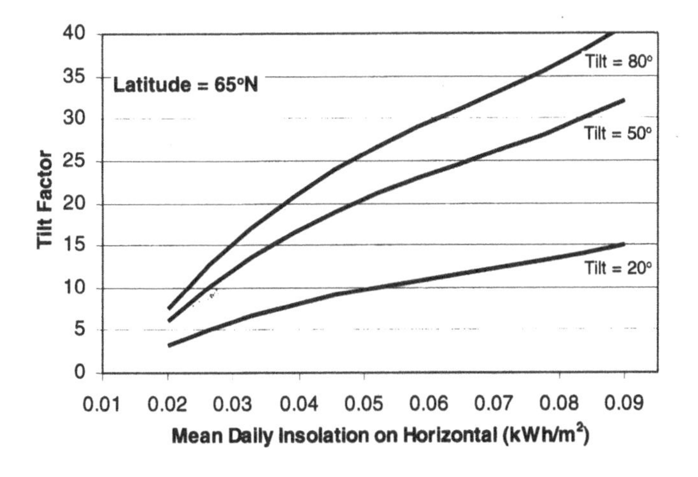

Figure A.5. Tilt factors for average daily radiation on the horizontal for December, 65°N latitude

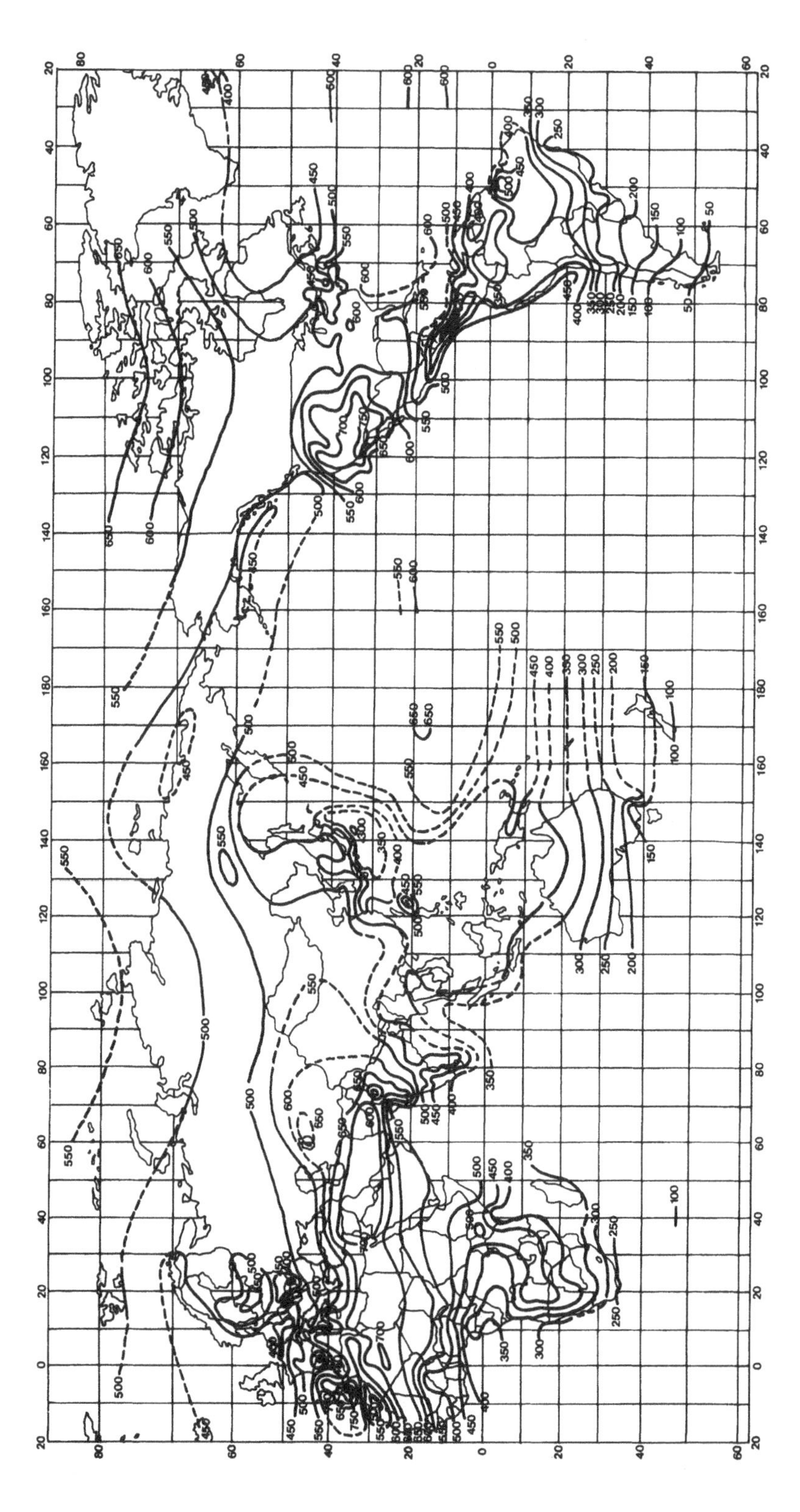

Figure A.6. Average daily radiation on a horizontal surface in June in cal/cm²
(map courtesy of Grundfos)

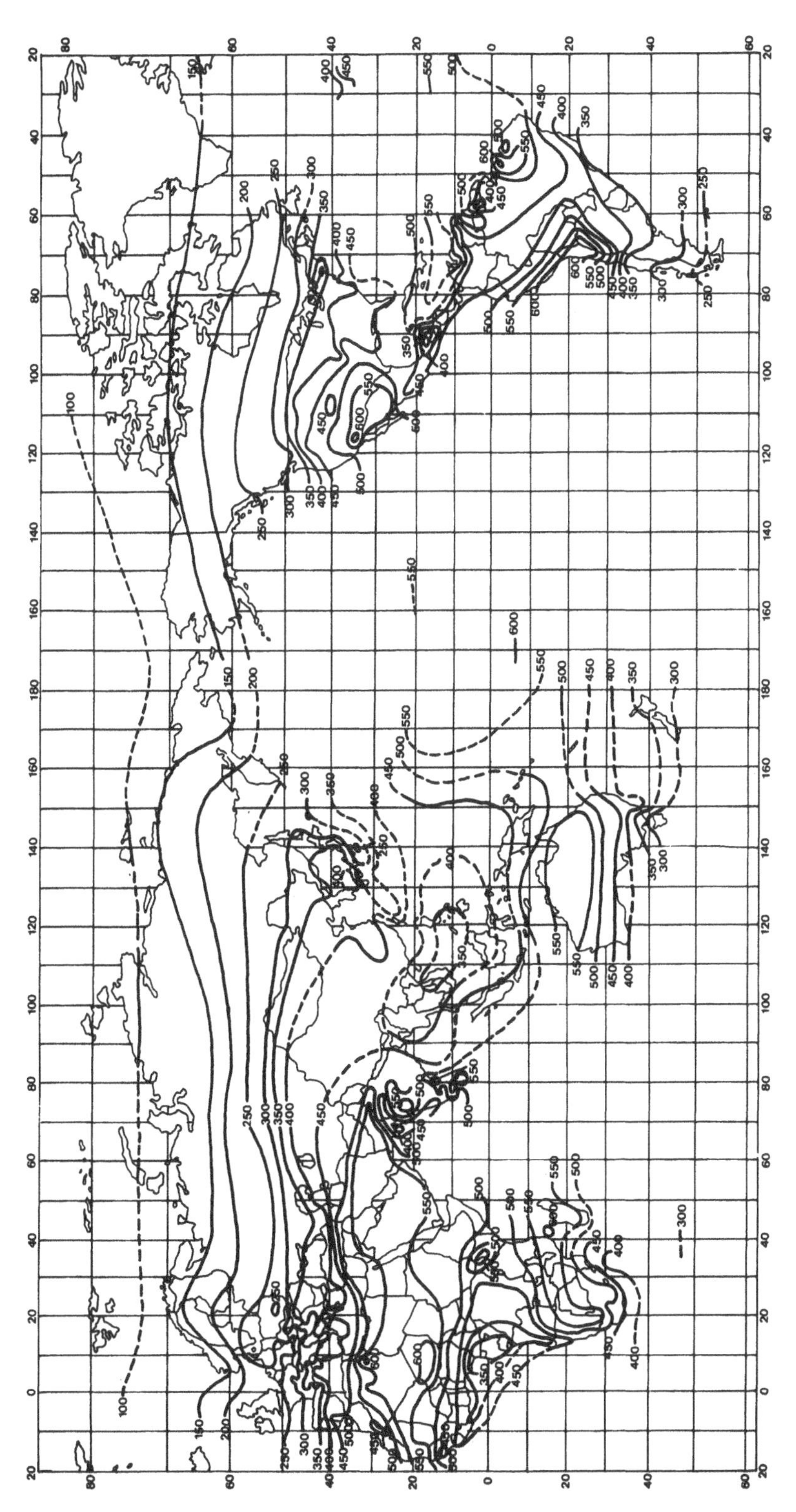

Figure A.7. Average daily radiation on a horizontal surface in September in cal/cm² (map courtesy of Grundfos)

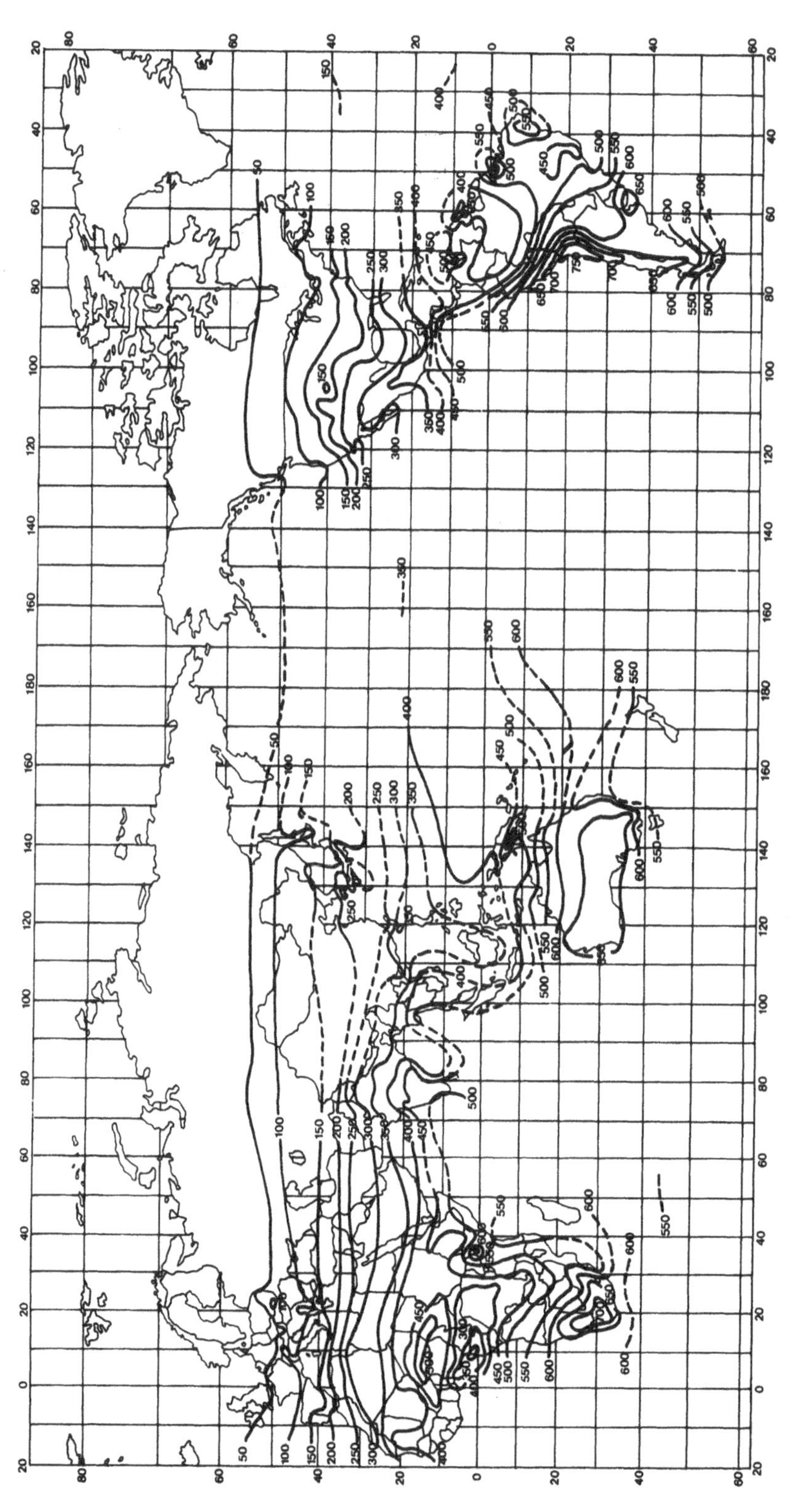

Figure A.8. Average daily radiation on a horizontal surface in December in cal/cm² (map courtesy of Grundfos)

B Effect of temperature on specific gravity readings

The specific gravity of a lead–acid battery during discharge can be used as a rough measure of its state of charge. It must be noted, however, that the specific gravity of sulphuric acid changes with temperature. Figure B.1 can be used to convert specific gravity measurements at temperatures between −30°C and 30°C to a standard temperature for comparison, that being 15°C.

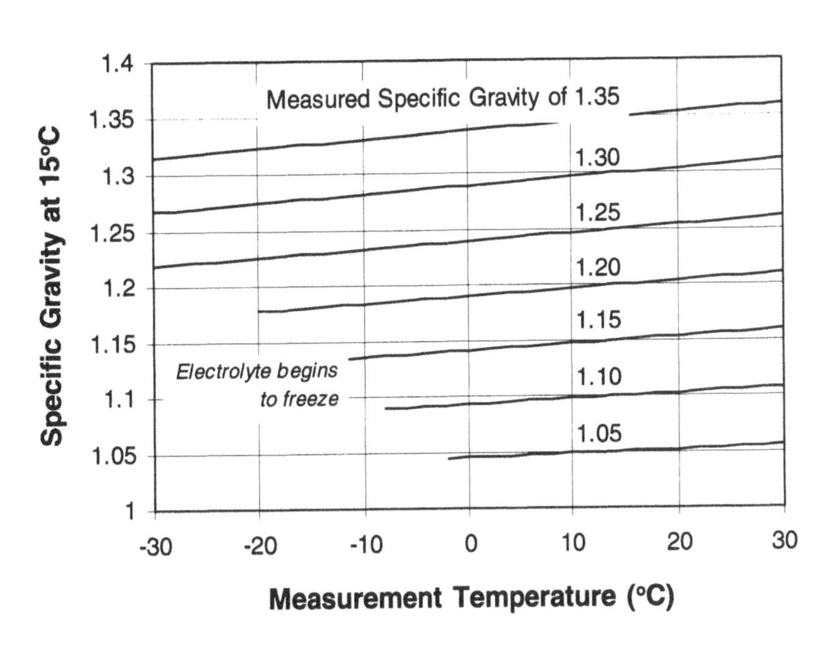

Figure B.1 Effect of temperature on the specific gravity of sulphuric acid

C PV component costs

The retail costs for the components found in PV systems are found below; these can be used in estimates of capital costs for PV systems, as discussed in Chapter 4.

MODULES

Usually the biggest single expenditure, *photovoltaic modules* were priced at an average of 4.15 USD per W_p in 1997 in North America.[1] By 2010, module costs should have dropped to 2–3 USD per W_p (see Figure 1.1 for past and projected module prices).[2]

BATTERY

In stand-alone systems, the *battery* is generally the second largest purchase. Lead–acid batteries are the most common type. Battery costs and lifetimes range widely, as shown in Table C.1.

BALANCE OF SYSTEM

The system cost also depends on the balance-of-system costs, i.e. the costs for *inverter*, *controller*, *array structure* (see

Table C.1 Approximate battery costs in 1998

Type of battery	Lifetime in cold climate (years)	Maintenance required	Cost per kWh* (USD) in North America
Lead–acid: automotive	1–3	Varies	50–80
Lead–acid: RV/golf-cart	2–4	Yes	70–100
Lead–acid: vented industrial	5–12	Yes	120–250
Lead–acid: inexpensive gel or AGM	3–6	No	160–250
Lead–acid: industrial gel or AGM	5–12	No	250–600
Nickel–cadmium: industrial	15–20	Yes	700–1250

* This is per kWh of battery capacity, not per kWh of energy stored in the battery over its lifetime.

Table C.2. Capital costs for inverter, structure and controller in 1998

	Inverter	Controls	Structure
Function	Converts DC current from battery to AC current for AC appliances	Regulates charging and discharging of battery	Mount for PV modules; some are fixed, while some track the sun
Requirement	Size of inverter must match maximum AC load power	Must handle maximum charge current	Size on basis of number of modules
Cost (USD)	0.50/W to 0.80/W for square-wave model; 0.80/W to 1.20/W for sine wave model (for sensitive equipment)	For arrays smaller than one kW_p, 30 to 200 More than 200 for larger arrays	Fixed: 0.50 to 1.00 per W_p of PVs Tracker: 0.75 to 3.00 per W_p of PVs
Notes	Prices are declining rapidly	Features affect price; sometimes inverter can perform charge control	Can also be integrated into buildings, permitting lower-cost mounting arrangements

Table C.3. Approximate battery enclosure costs in 1998

Type of battery enclosure	Cost per kWh of battery capacity (USD)
Uninsulated: plywood, plastic aluminium or steel	5–50
Insulated	60–120
Insulated with water as phase change material	100–200

Table C.2) and *battery enclosure* (Table C.3). In cold climates, it makes sense to insulate the battery enclosure: if the batteries are protected from the cold a smaller battery bank can be used.[3] Expensive batteries are sometimes accompanied by a large mass of water. This maintains battery temperatures near 0°C, permitting significant reductions in battery size.

REFERENCES

1. Maycock P (1998). U.S. PV Cell/Module Shipments Increase 36.2 Percent. *Photovoltaic News* (February).
2. Leng G, Dignard-Bailey L, Bragagnolo J, Tamizhmani G, Usher E (1996). *Overview of the Worldwide Photovoltaic Industry*. Varennes, Québec: CANMET Energy Diversification Research Laboratory.
3. Ross M, Egles D (1997). Valuing the Benefits of Insulated Battery Enclosures for Stand-alone Photovoltaic Systems in Cold Climates. *Proceedings of the 23rd Annual Conference of the Solar Energy Society of Canada, Inc.* Ottawa: SESCI.

D List of contributors

Clément Bergeron
Environergie Inc.
561 rue Pacifique
Lac-St-Charles
Quebec G3G 1W6
Canada
Tel. +1 418 849 5221
Fax +1 418 849 5221
Email envirorj@clic.net

Dr Alf Bjørseth
ScanWafer AS
PO Box 290
8160 Glomfjord
Norway
Email alf.bjorseth@scanwafer.com

Jito Coleman
Northern Power Systems
One North Wind Road, PO Box 999
Waitsfield
VT 05673-0999
USA
Tel. +1 802 496 2955
Fax +1 802 496 2953
Email jcoleman@northernpower.com
Web page http://www.northernpower.com

Dave Egles
Soltek Solar Energy Ltd
2-745 Vanalman Avenue
Victoria, BC, V8Z 3B6
Canada
Tel. +1 250 727 7720
Fax +1 250 727 2135
Email soltek@soltek.ca
Web page http://vvv.com/~soltek

Viggo Gahrn
Tele Greenland A/S
Thoravej 4,
DK-2400 Copenhagen NV
Denmark
Tel. +45 38 34 22 44
Fax +45 38 34 63 65

Dr Heinrich Häberlin
Berner Fachhochschule,
Hochschule für Technik und Architektur Burgdorf
Jicoweg 1
CH-3400 Burgdorf
Switzerland
Tel. +41 34 426 68 53
Fax +41 34 426 68 13

Alf Hedman
Copyskylt AB
Box 105, S-26121
Billeberga
Sweden
Tel. +46 418 4314 00
Fax +46 418 4313 85

Knut Hofstad
Norwegian Water Resources and Energy Administration
Middelthunsgate 29,
Postboks 5091 Maj.
0301 Oslo
Norway
Email KHO@nve.no
Web page http://www.nve.no

Ing. Josef Huber
Tauernkraft
Rainerstrasse 29
A-5020 Salzburg
Austria
Tel. +43 662 8682 22285
Fax +43 662 8682 12 22285
Email huberjos2@verbund.co.at

Thomas A. Lawand and Joseph Ayoub
Brace Research Institute
McGill University
PO Box 900 Macdonald Campus
Ste-Anne de Bellevue
Quebec H9X 3V9
Canada
Tel. +1 514 398 7833
Fax +1 514 398 7767
Email ae12@musica.mcgill.ca

Dr Peter Lund
Helsinki University of Technology
PO Box 2200
FIN-02015 HUT
Finland
Tel. +358 9 451 3197
Fax +358 9 451 3195
Email peter.lund@hut.fi
Web page http://www.hut.fi

Sylvain Martel
CANMET Energy Diversification Research Laboratory, Natural Resources Canada
1615, boul. Lionel-Boulet
Varennes, Quebec J3X 1S6
Canada
Tel. +1 450 652 6747
Fax +1 450 652 5177
Email sylvain.martel@nrcan.gc.ca
Web page http://cedrl.mets.nrcan.gc.ca

Dr Bengt Perers
Vattenfall Utveckling AB
Box 1046,
S-611 29 Nyköping
Sweden
Fax +46 155 293060
Email bengt.perers@utveckling.vattenfall.se
Web page http://www.vattenfall.se

Chuck Price
Sovran Energy Inc.
13187 Trewhitt Road
Oyama, B.C. V4V 2B1
Canada
Tel. +1 250 548 3642
Fax +1 250 548 3610
Email crprice@infoserve.net

Michael Ross
(formerly of the CANMET Energy Diversification Research Laboratory, Natural Resources Canada)
Email habib_ross@hotmail.com

Jimmy Royer
Solener Inc.
442, rue Lavigueur
Quebec, Quebec G1R 1B5,
Canada
Tel. +1 418 640 7444
Fax +1 418 640 7445
Email Solener@gbc.clic.net

Dipl.-Ing. Dr Gerd Schauer
Verbundgesellschaft
Am Hof 6a
A-1010 Wien
Austria
Tel. +43 1 53113 52439
Fax +43 1 53113 52469
Email SchauerG@verbund.co.at

David Spiers
Neste Advanced Power Systems UK
PO Box 83, Abingdon
Oxon OX14 2TB
UK
Fax +44 1235 553 450
Email David.Spiers@neste.com
Web page http://www.neste.com/group/energy/naps/index.html

Fritjof Salvesen
KanEnergi AS
Bærumsveien 473
1352 Rud
Norway
Email kanenergi@kanenergi.no

Dr Raye Thomas
NewSun Technologies Ltd
74 Auriga Drive #2
Nepean, Ontario K2E 7X7
Canada
Tel. +1 613 723 5750
Fax +1 613 723 5980
Email rthomas@newsun.ca

Dr Juha Vanhanen (formerly of Helsinki University of Technology)
Otaniemi Consulting Group Oy
Lönnrotinkatu 19 B
FIN-00120 Helsinki
Finland
Tel. +358 9 6866 6230
Email juha.vanhanen@kolumbus.fi

Dipl.-Ing. Heinrich Wilk
Oberösterreichische Kraftwerke AG
Böhmerwaldstrasse 3
A-4020 Linz
Austria
Tel. +43 732 6593 3514
Fax +43 732 6593 3309
Email heinrich.wilk@oka.co.at

Index